Jesús Velázquez Macias
Claudia Guadalupe Lara Torres
José Alberto Vela Dávila

A Telegram Bot for the Treatment of Smoking Cessation

Jesús Velázquez Macias
Claudia Guadalupe Lara Torres
José Alberto Vela Dávila

A Telegram Bot for the Treatment of Smoking Cessation

Using a Telegram Bot in the Treatment of Tobacco Dependence

ScienciaScripts

Imprint

Cover image: www.ingimage.com

This book is a translation from the original published under ISBN 978-613-9-40584-8.

Publisher:
Sciencia Scripts
is a trademark of
Dodo Books Indian Ocean Ltd. and OmniScriptum S.R.L publishing group

120 High Road, East Finchley, London, N2 9ED, United Kingdom
Str. Armeneasca 28/1, office 1, Chisinau MD-2012, Republic of Moldova, Europe
Managing Directors: Ieva Konstantinova, Victoria Ursu
info@omniscriptum.com

Printed at: see last page
ISBN: 978-620-8-59524-1

USE OF A TELEGRAM BOT IN SMOKING CESSATION TREATMENT

"IMPLEMENTATION OF A TELEGRAM BOT FOR MOBILE DEVICES TO SUPPORT SMOKING TREATMENT "

JESÚS VELÁZQUEZ MACIAS
CLAUDIA GUADALUPE LARA TORRES
JOSÉ ALBERTO VELA DÁVILA
Polytechnic University of Zacatecas
TecNM: Higher Technological Institute of Fresnillo

Summary

The study addresses the implementation of a chatbot aimed at managing smoking cessation treatment, carried out in an Integration Center for Youth with patients in active treatment for smoking cessation. The main objective was to evaluate the impact of a Bot on Telegram on treatment management, replacing traditional methods and streamlining registration and consultation tasks, through the use of freely accessible technological platforms. The hypothesis is that this approach would increase treatment effectiveness. The research, quantitative in nature, employed an exploratory cross-sectional non-experimental design and a pre-experimental experimental approach, with an exploratory and correlational scope. Questionnaires and parametric tests were used for data collection. The results showed that the variables presented a low correlation, indicating that the use of Bot did not significantly improve the reduction of consumption according to the therapists. However, it was concluded that the tool and the web platform that supported it significantly facilitated treatment management, optimizing the time and resources of health professionals during the study.

Keywords: smoking, chatBot, impact on treatment, treatment management,

Inde

INTRODUCTION

The use of mobile technologies has been extended to countless areas of research in the human sciences, providing tools that simplify or complement their work, in the case of health sciences there have been great contributions that benefit both health specialists and their patients, In the case of smoking, treatments require records and logs made by patients to be analyzed and interpreted by therapists trained to treat this type of addictions, who determine the best treatment based on these records and behaviors.

The functions provided by a mobile application and integrated with an instant messaging client provide the ideal platform to manage and support smoking treatment related tasks, all through a digital process.

In this sense, the general objective of this study is to "Determine the impact of the use of a Telegram Bot for the management of patients in treatment with smoking", in order to record activities on the patient's side, related to tobacco consumption in addition to having the functionality to receive health advice, reminders of appointments and therapies.

The tool and the research design were developed following the specifications and attending the methodologies of attention and prevention of addictions provided through the documentation and literature of a Integration Center for young people in .the city of Zacatecas

The following statement is presented as a general hypothesis; "The use of a Telegram Bot increases the impact on the management of smoking cessation treatment", and to accept or discard that premise a methodology based on a quantitative approach with an exploratory and correlational scope and an experimental design of the pre-experimental type was used, in addition to another non-experimental design of the exploratory cross-sectional type.

This research is considered to meet the requirements that frame the research line "e-Health" and is focused on the sub-type "Mobile Health" because by means of a mobile application those involved in this study will collect and manage the information in support of the treatment against smoking to subsequently interpret and analyze the results obtained, some years this type of applications and the respective technology that supports them, there have been several contributions that highlight the importance of this type of technologies in support of health care in different areas.

Therefore, as argued in previous paragraphs, this paper mentions the development carried out during the entire researchdetailing each of its main aspects through the following sections, which are summarized below;

Chapter 1 deals with the problem statement, in which the issue or question to be investigated is presented, including its theoretical bases, objectives and background, and mainly the incidence related to the use of technologies related to chat bots for mobile devices in the treatment against smoking, defining and structuring in a formal way the idea that arises from this research.

Chapter 2 contains the assertions related to the theoretical framework related to mobile applications that intervene in health-related processes, their background and results obtained. In addition, this part is made up of several elements, including: a collection of references, theoretical concepts and background on which the research is based, its state of the art, and some theories directly related to this research.

Chapter 3 describes the methodology used in the development of the study, presenting and describing the main design processes of the research, specifying the phases that were carried out, explaining the methods used to obtain information and the instruments that were used and designed for this purpose.

Chapter 4 reports on the results obtained; this section is used to describe the analysis of the data obtained through the different instruments designed. A total of 4 instruments make up this section. At the beginning of the analysis, the instrument designed for is used to determine the acceptance or rejection of the patients by means of the "Motives for consumption" questionnaire. Subsequently, usability analyses of both the Bot and the web platform that supports the therapists are carried out, and finally the results obtained from the "Effectiveness in treatment" questionnaire are examined to determine the behavior of the consumption rates by the patients.

Finally, Chapter 5 presents the conclusions and discussions. This chapter begins by verifying the hypotheses presented and then concludes with questions related to the data obtained from the analysis of Chapter 4 labeled as results. Some benefits and areas of opportunity of the tool used, in this case the Bot, are established. Finally, some guidelines are left open to research where the need to review and update some criteria that allow the best use of this type of tools for human health is exposed.

It is also important to clarify that the project turned out to be feasible even with the limitations and restrictions presented for the development of the study within the juvenile integration center, due to the norms established in the handling of information and respect for the privacy of the patients, as well as the resistance to technology expressed by the therapists and patients.

CHAPTER I PROBLEM STATEMENT

This section sets out the rationale of the object to be studied in this document, its theoretical bases, objectives and background, but mainly the incidence related to the use of technologies related to chat Bots for mobile devices in the treatment against smoking, defining and structuring in a formal way the idea that arises from the present research, as stated in the title of this section, raising the problem is defined as the action of refining, specifying and structuring the research idea (Hernández-Sampieri and Mendoza Torres, 2018)

Going deeper into the subject, the problem statement is the first step when talking about scientific research, observing the context of the study, from the problematic entity arises the research formulation, the methodology and even the title of the thesis (Arias Gonzáles and Covinos Gallardo, 2021) .

Given the magnitude of this important step in the research, Hernández-Sampieri and Mendoza-Torres (2018) mention that "a correctly posed problem is partially solved; the greater the accuracy the more possibilities of obtaining a satisfactory solution" (p. 40), therefore, each of its elements should be supported by specialized literature on this subject to have a firmer and more orderly start, with a solid foundation the definition of the problem will be the instrument to define more precisely the objectives, content and procedure of study (Baena Paz, 2017) .

The organization of the information facilitates the distribution of tasks and clarifies the author's work by ordering the elements according to the position they occupy within the research. For this reason, the structure of this research document is based on the hexagon method, proposed by Arias González and Covinos Gallardo (2021), in which the problem statement is divided into the following sections as shown in Figure 1.

Figure1 *Schematization of the development of a problem statement*

Note: Adapted from Esquematización del proceso de elaboración de un planteamiento del problema, by Arias Gonzales and Covinos Gallardo (2021).

The following is a brief description of each section and its analogy with the sections of this document; in the first place, the problem situation consists of describing what is observed as something that is to be investigated because it is new or attractive to the researcher, . This section is included in the section on the basis of the research problem.

The second box talks about the theorization of the phenomenon of the study, it has to do with the topic of study in question and with the authors that validate and support this concept and are also in tune with the context of the study itself, with current statements, likewise the research background box is related to the selection of recently conducted studies that address concepts related to our own research, both boxes are developed in this document during the stages of description of the problem and more in depth within the chapter of the Theoretical Framework.

The boxes on the importance of the study and the author's purpose refer to the justification of the research, where the authors explain the reasons why the study was conducted, as well as the benefits and achievements obtained at the end of the research, here the authors explain the reasons for conducting the research or study (Arias Gonzáles and Covinos Gallardo, 2021) .

The general problem is described and assimilated as the research questions proposed in this document, which are the formal expression of the problems that the research wants to solve , delving into the theory of the phenomena of interest, analysis of previous studies, expert opinions, among others.

1.1 Description of the specific research problem.

The idea for the development of this research arises after the development of the article "Use of a Bot for the verification of mathematical formulas of the subjects of Probability and Operations Research at the Polytechnic University of Zacatecas" (Velázquez Macías et al., 2016) , which describes a tool developed for mobile devices supported by the instant communication platform Telegram, which is available to verify calculations of Probability and Operations Research, students can corroborate the results of their exercises and in turn teachers can review booklets more quickly.

In the mentioned article, times with and without the tool are presented, being all of them less than the traditional way of checking and reviewing, all this using an assistant that simulates to be a person and with which you can "chat" to obtain certain results, this type of interaction and the technology behind this type of program is known as Bot, currently the Bots are used as help or support systems before physical persons, with this the response times are reduced and easy queries are attended without the need for a person to intervene (Chesñevar and Estevez, 2018) .

A Bot is considered a software application added with a contact to an instant messaging service that interacts with web services and its responses are based on knowledge bases and represented by natural language text messages

At the Juvenile Integration Center of Zacatecas there are a large number of forms to be filled out by a patient under treatment for smoking, which consist of logs and consumption records, this task is done physically on sheets of paper which are easily forgotten, lost, etc., this slows down the treatment and prevents therapists from proposing an effective treatment, a Bot application could be used to record the patient's activity by having the device always at hand, we hardly leave home without the cell phone.

Through the use of a Bot, the statistical information of each patient would be available to therapists in databases hosted on the Internet, which would help to provide more timely treatment, taking into account the benefits of this type of assistant, the savings in paperwork to fill out the forms, and the ease of recording events, or receiving notifications, consequently and based on the experience gained in the development of the Bot described above, the idea of developing a second Bot to manage the processes of treatment against smoking was proposed.

After some talks with the staff involved in these processes the idea was proposed by the researcher and listing the benefits of a personal assistant based on instant messaging could help manage treatment processes, the design and development of this tool was expressed with the publication of the article called "Development of a Bot for support in the treatment of smoking in the Juvenile Integration Center in Zacatecas" (Velázquez Macías et al., 2017) , said tool meant something novel in 2017, both for the staff involved and for the treatment itself, since in that year Bots were still not as popular as they are today.

That is why in this research work we intend to demonstrate the reliability of the use of a Telegram Bot considered as a personal assistant for mobile devices (Mulyanto, 2020) , by the way it can interpret various commands to perform consequent actions, in addition the advantages of its implementation over the traditional method are highlighted.

Based on the above information, the following situations described in sections 1.1.1 and 1.1.1.2 can be concluded, referring to the traditional work method, i.e. how the work is carried out prior to the use of Bot and the desirable situations planned during its implementation.

1.1.1 Current situation

The patient in smoking cessation treatment fills out physical forms by hand, such as logbooks, which must be taken to the next session with the therapist, but many times the patient forgets or misplaces them, the therapist has incomplete information to give his diagnosis, thus lengthening and slowing down the treatment, by forgetting or misplacing the forms the patient loses important follow-up information, because of this there is no report on the level of timely progress in the treatment of each patient.

The time spent on care is reduced to trying to recover lost information and the essential part of monitoring progress to establish an accurate diagnosis is neglected, and patient history files are kept in physical format.

1.1.2 Desirable situation

The Bot is added as a contact in the Telegram mobile application previously installed on their mobile device, after that, the patient in anti-smoking treatment keeps their records and smoking habits interacting with the Bot.

The patient provides his or her unique identifier and the therapist accesses the database that Bot previously generated, through the web module designed to perform queries and extract information, the therapist visualizes the patient's records and habits, which allows him or her to make an accurate diagnosis and follow-up, in addition the patient can receive periodic notifications about important information related to his or her treatment, even if the patient loses or changes his or her cell phone or forgets it the day of the appointment, the therapist can recover his or her records, thus speeding up the management of the patient's treatment.

1.2 Grounding of the research problem (Conceptual Background)

Smoking according to the World Health Organization (WHO) ranks second among the leading causes of death, only in 2015 caused approximately 11.5% of total mortality, with a total estimate of 1.1 billion smokers around the world, snuff produces serious health effects such as cancer, cardiovascular and respiratory diseases (Yoong et al., 2020) . Tobacco is a risk factor for the four major non-communicable diseases (NCDs): cancer, cardiovascular diseases, diabetes and respiratory diseases (Blanco et al., 2017) .

In 2008, the WHO created the MPOWER package, a technical tool that includes more effective tobacco control measures to halt the tobacco epidemic. These measures are intended to protect the population from the use and negative consequences of tobacco;

- Monitoring and surveillance of tobacco use and prevention measures
- Protect the population from exposure to tobacco smoke
- Offer assistance in smoking cessation
- Warn about the dangers of tobacco
- Enforce prohibitions on tobacco advertising, promotion and sponsorship
- Increase tobacco taxes

To support the above international regulations, the consumption and distribution of tobacco in Mexico is regulated by law and the treatments related to its effects are endorsed by General Law for Tobacco Control (2010) which in its Article 9 mentions that;

> The Secretariat will coordinate the actions developed against smoking, promote and organize early detection services, guidance and care for smokers who wish to quit, investigate its causes and consequences, promote health considering the promotion of attitudes and behaviors that favor healthy lifestyles in the family, work and community; and develop permanent actions to dissuade and prevent the consumption of tobacco products mainly by children, adolescents and vulnerable groups. (p. 4)

In the preceding paragraph, the term "secretariat" refers to the secretariat of health of the United Mexican States.

In response to the international call, Mexico was the first Latin American country to confirm its participation in the World Health Organization's Framework Convention on Tobacco Control (FCTC) in 2004. This convention relies entirely on the parameters and metrics established by the MPOWER indicators, Mexico has made significant progress in controlling the tobacco epidemic, however, for the whole movement to work, the full implementation of the FCTC is necessary to be able to offer effective smoking cessation treatment (Zavala-Arciniega et al., 2019) .

Statistically speaking, there are about 11 million active smokers in Mexico and smoking is responsible for 60 thousand deaths per year due to tobacco smoke inhalation . For this reason, youth integration centers give priority to people addicted to tobacco with the help of therapists and psychologists who use various techniques to help them control their addiction

As in Mexico, in the state of Zacatecas alcohol and tobacco consumption generate a public health problem that is constantly increasing, especially in adolescents, consumption is higher than the national average and the age at which they start consumption is increasingly lower (Legaspi et al., 2020) .

In the National Survey on Drug, Alcohol and Tobacco Consumption (ENCODAT) 2016-2017 in its Tobacco Report edition, it maintains that 195 thousand Zacatecans are active smokers (37 thousand women, 158 thousand men), of which 84 thousand smoke daily and 111 thousand smoke casually (Instituto Nacional de Psiquiatría Ramón de la Fuente Muñiz, 2017) , in relation to the indicators established by MPOWER the national survey reflects the following results is its different categories in its global sections for the state of Zacatecas:

- Monitor, in Zacatecas 18% of the population smokes tobacco, the average age of initiation of daily tobacco use is 21.9 years in women and 18.0 years in men.
- Protect, in Zacatecas the public places marked with the highest prevalence of secondhand tobacco smoke by non-smokers are: bars, restaurants, public transportation and schools.
- Offering help, 80.1% of current smokers in Zacatecas are interested in quitting smoking in the future.
- Among active smokers in Zacatecas, 45.4% think about quitting smoking because of the illustrated health warnings.
- During the last month, 77.8% of the population of Zacatecas watched or listened to information on television and radio about the risks of smoking.
- Taxes, among active smokers in Zacatecas 53.8% bought cigarettes per unit, 72.4% of the population supports the tobacco tax increase law.

In other results, ENCODAT states that Zacatecas ranks tenth in the Mexican Republic in terms of smoking prevalence, and is also one of the three least prevalent states in terms of second-hand tobacco smoke in schools, and also ranks first in terms of the use of health warnings with illustrations to quit smoking.

For this thesis, and having all the background on the importance of managing a treatment against smoking efficiently and given the importance of this serious health problem, it was proposed to create a chat Bot or Bot, which works as a support to the treatment already established and validated in Juvenile Integration Centers in Zacatecas.

Centros de Integración Juvenil is an institution focused on prevention, treatment, rehabilitation, scientific research and creation of specialists in drug use issues, in its web page and as main objective, Centros de Integración Juvenil (n. d.) mentions:

> Contribute to mental health and drug demand reduction with the participation of the community through prevention and treatment programs, with gender equity, based on evidence to improve the quality of life of the population. (p. 9)

In addition, its mission is to provide mental health prevention and treatment services to mitigate drug use, with judgments of equity, equality and non-discrimination, based on scientific knowledge and specialized professional staff.

After analyzing the bibliographic material according to the topic of mobile applications related to medical issues and more specifically applications related to addiction treatment, an initial approach was made to some projects related to this type of technology associated with human health, being similar to the current research project , some theories correlated with its operation that directly affect its application success will be taken as a reference.

1.3 Delimitation of the problem

The delimitation of the problem is done in terms of time, i.e. when it happened, on what dates, in what space, what are its antecedents, which originated it (Baena Paz, 2017) , therefore the delimitation of the present work is of vital importance to define the actions associated with the research without exceeding the proposed limits and covering everything necessary to satisfactorily fulfill the work, in other words, Arias Gonzáles and Covinos Gallardo (2021), argue that the scopes are based on stating how far the researcher wants to go with his study.

The present research was developed during the period comprising the years 2017 to 2018 at the Centro de Integración Juvenil Zacatecas, with address at Parque Magdaleno Lujan s/n, Colonia Buenos Aires in the city of Zacatecas, Zacatecas, Mexico, Located northeast of the city.

The study included a population of patients of legal age undergoing treatment for smoking divided into 2 groups, one of 28 and the second of 32, and also included the therapists or health professionals in charge of managing the treatment, associated personnel, commissioners, volunteers and directors of the institution.

1.4 Research Questions.

As central questions the research questions help to raise what is planned to be answered through studies or research, to clarify the problem can work the development of several research questions, which should have as far as possible certain characteristics such as clarity, feasibility and relevance (Baena Paz, 2017) .

To better clarify a problem, the elaboration of several research questions works, which are the articulation of the researcher's ideas that contemplate a relationship between the variables found for further analysis (Baena Paz, 2017) .

It is difficult to express in one or several questions the problem in a total way taking into consideration all its content, many times only purpose of the study is stated, in any sense the questions should synthesize what will be the research (Hernández-Sampieri et al., 2018) , taking into account the recommendations and previous aspects, in the following sections 1.4.1 and 1.4.2, the general and secondary research questions related to this research work are defined.

1.4.1 General Research Question.

PG. What is the impact of using a Telegram Bot for tobacco treatment management?

1.4.2 Secondary Research Questions.

Q1. How to assess in an automated way the viability of patients to verify if they are suitable to undergo anti-smoking treatment?

Q2. What is the acceptance rate of using a Bot for mobile devices for tobacco treatment management?

Q3. How effective will anti-smoking treatment be with the integration of a Bot for mobile devices?

Q4. What is the acceptance rate using a web platform for tobacco treatment management?

1.5 General Objective

To specify the tasks of any research, the objectives are written to set the line to be followed by the researcher, or in other words to define what you want to accomplish (Arias Gonzales and Covinos Gallardo, 2021) , with this is intended to contribute to the resolution of a problem with the drafting of clearly expressed, specific, measurable, appropriate and realistic objectives (Hernandez-Sampieri et al., 2018)

It is also important to mention how the research is thought to help solve this problem, the objectives may determine limits on the research that usually should be achievable, for different reasons sometimes this is not possible due to lack of resources or time impossibility (Baena Paz, 2017) .

Complementing the previous statements, the general objective responds to the general research question, the wording of the text is similar, the main difference is that it does not include questions and an infinitive verb must be used at the beginning of the sentence (Arias Gonzáles and Covinos Gallardo, 2021)

addition, the general objective must meet certain attributes such as: a) Qualitative, highlighting the quality of the research, b) Integral, since it integrates at least two specific objectives and c) Terminal, when reaching the goal ends, it is not permanent, i.e. the general objective is reached only once (Caballero, 2014) , having said this and continuing with the description process as far as the present research is concerned, the general objective would be as follows:

OG. "Determining the impact of using a Telegram Bot for smoking cessation treatment management

1.5.1 Specific objectives

The specific objectives are the achievements that the researcher wishes to obtain in order to reach the general objective. These achievements can be sequential or parallel; therefore, they can be proposed as they are achieved in chronological order or at the same time (Arias Gonzáles and Covinos Gallardo, 2021) , in view of these statements, the specific objectives contemplated in the research of this document are detailed below

SO1. "Design a scenario that allows automated assessment of the viability of patients to verify whether they are fit to undergo anti-smoking treatment".

SO2. "Determine the rate of acceptance of the use of a Bot for mobile devices for the management of smoking cessation treatment".

SO3. "To evaluate the efficacy of anti-smoking treatment by integrating a Bot for mobile devices".

SO4. "Determine the acceptance rate of the web platform for the management of smoking cessation treatment".

1.6 Justification

The justification is the part of a research project that states the reasons that motivate the author to initiate such research, in other words, what was the researcher's need to select the topic to develop it (Baena Paz, 2017) , the vast majority of research is conducted with a defined purpose in mind, that purpose must be sufficiently representative to justify its realization (Hernández-Sampieri et al., 2018) .

Some authors recommend answering some questions to support the writing of this part of the researchthese questions should be aimed at knowing the main aspects of the study, for example, Caballero (2014) proposes answering the following questions for a correct writing of the justification:

a) For whom is this research necessary?

b) Why is it done?

c) For whom is it convenient?

In response to the above recommendations and in order to answer the question "why?", the following arguments are presented: the thesis topic was chosen as a contribution to the solution of a current health problem such as smoking. Relying on Information Technologies and the development of applications for mobile devices, the use of technology in medical issues can be adopted as a support or platform to perform more efficiently the work related to human health, when developing this type of tools, some aspects of design, connectivity and access to technology must be taken into account.

The importance of this research lies in facilitating treatment management both for patients under active treatment and for the therapists or health professionals assigned to them, providing them with technological tools that streamline the processes of records and information queries, in order to achieve a more efficient interpretation of the data and reach a follow-up diagnosis of each patient and provide the necessary care and recommendations to successfully reach the end of the treatment.

This research is necessary to improve the optimization of resources, replacing paper records, logs or queries with their equivalent in digital version, thus avoiding the loss of information, the agility in searches, the organization of information and the care for the environment.

It is convenient for patients, who must comply with certain requirements during their treatment, as the main part, to keep records of daily tobacco consumption, the fact of being able to do it through their cell phone and by means of the personal assistant is more practical and avoids bringing with them the physical record on sheets of paper.

It is also convenient for the Centro de Integración Juvenil Zacatecas to improve its administrative processes, such as customer service, treatment management and general administration of patient files when undergoing active treatment, in this case related to smoking.

Another important point is that, since it does not require any type of resource, sponsorship or economic support, nor the acquisition of software licenses or computer equipment to carry out the project, it does not imply for the institution or the researcher to have to disburse any type of economic resource , since free software and computer equipment available in the institution were used to carry out the project, so the realization of the project does not depend to a greater extent on economic resources.

1.7 Feasibility of the research

The determination of whether a research is feasible or not, is equally important as the objectives, questions, and justification of the research which were described in previous paragraphs, this viability or feasibility of the research, must consider aspects related to knowledge and skills needed to develop the project, availability of time, financial resources, human resources and material resources that will define, the scopes of the research (Mertens, 2019)

Having said the above, it should be questioned whether realistically there is access to each and every one of the actors related to the project and the consequent information (Hernández-Sampieri and Mendoza Torres, 2018) , in such a way that the completion of the research is not at risk due to factors external to it that are not foreseen and that negatively influence the natural development of the project

On this basis, it was possible to determine that the present research is feasible since we had the institutional support and the support of the management staff of the Juvenile Integration Centers of Zacatecas regarding the use of the proposed tools such as the Telegram Bot and the Web platform for consultation by the therapists, in addition to providing the researcher with computer resources, facilities and furniture necessary for the realization of the project.

CHAPTER II THEORETICAL FRAMEWORK

This chapter has several elements among which the following stand out: a collection of references, theoretical concepts and background on which the research is based, in general the theoretical basis of the project, as well as its state of the art, complemented by some theories directly related to the research issues, and a compendium of terms related to the research, which in the opinion of the author of this document are important to introduce the reader who does not belong to the area of study, to understand what is being referred to with the technical terms, in each of the sections of this research.

As a first point, Arias González and Covinos Gallardo (2021), consider reviewing the existing literature from the research topic is considered valuable in the sense that it will help to form the theoretical roots of the study. In 2017, Baena Paz argued that "knowing the above is a requirement for the selection of a research problem" (p. 96), therefore, identifying the lines of research previously worked, their methods and techniques support the complementation of the topic, the clarification of the research problem as a continuity to the previous works found.

Hernández-Sampieri et al., (2018) define literature review works as; "detecting, consulting and obtaining the bibliography (references) and other materials that are useful for the purposes of the study, from where has to extract and compile the relevant and necessary information to frame our research problem" (p. 61), this review has to be selective due to the immense amount of information that could be found.

In 2018, Hernández-Sampieri et al. coined the term "vertebrate" in reference to the construction of a theoretical framework by making a global or general tentative index to then refine it until it is very specific, to then place the bibliography in its corresponding places, the vertebrate process consists of defining topics and subtopics to then associate the references to one or more subtopics, as can be seen in Figure 2.

Figure2 *Method of vertebration of the theoretical framework index and assignment of references*

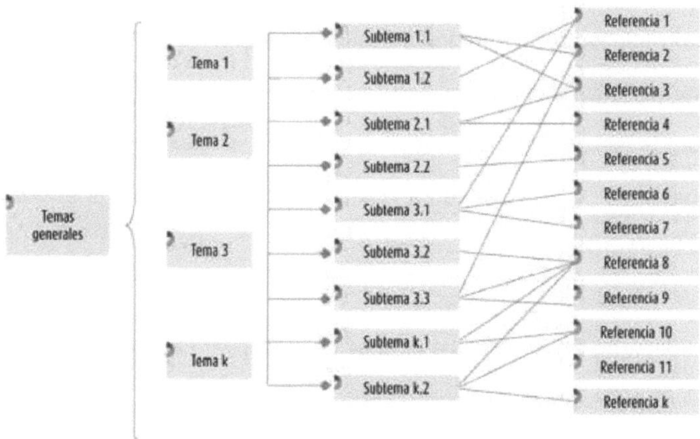

Note. taken from (Hernández-Sampieri et al., 2018, p.x. 79).

The above method allows structuring the theoretical framework to organize hierarchically the topics and subtopics as well as assigning the bibliography to one, two or more topics as appropriate, in other words in this chapter are collected some of the most recent publications directly related to the subject of this research specifically on mobile devices to address issues related to the care and preservation of human health, an additional and complementary section will address the technical concepts related to this research and that in one way or another involved either directly or indirectly with the development of this work.

2.1 State of the art

Also called by some authors as state of the question; it reveals in a summarized way the level of progress of some work, this progress will contain references and hypotheses that give an example of the knowledge of the subject in relation to the progress of the researcher (Baena Paz, 2017) , in other words, Hernández-Sampieri et al., (2018), argues that the state of the question, is "knowing the current situation of the problematic, what is known and what is not, what is written and what is not written, what is evident and what is tacit" (p. 466).

Following the trend of the arguments put forward by the authors cited in the previous paragraph, following sections present various research studies related to the one discussed in this document as part of the analysis of the state of the question.

2.1.1 Information technology in the service of health care

There are different studies that attest to the different types of information technology contributions in favor of health, these studies are reflected as articles in scientific journals or as graduate or postgraduate theses, these investigations have had a positive impact on the work of health personnel in terms of prevention, information or treatment, perhaps some to a greater or lesser extent, proof of this is the large number of mobile applications and chat Bot that arise from these studies and that stand out in this area, the following are the most relevant in the personal opinion of the author of this paper.

Some of these applications intervene, for example, with the management and administration of medical records and monitoring of physical and physiological variables in children and adolescents (Grimaldo Botero, 2013) , others promote the oral health of patients through interactive content and illustrations (Anderson and Montero, 2021) , different studies analyze the effectiveness of mobile applications by categorizing and ordering them according to their function, issuing at the end a critical opinion of each of them (Briz Ponce, 2016) , in general, these types of studies are related to the one proposed in this research, because they involve medical issues in their procedures supported by some type of mobile application for their purpose.

In this same category there are abundant studies on medical applications related to the taking of medicines that analyze the advantages and disadvantages of each one of them (Fernández, 2020) , others analyze statistical data on the use of mobile applications and how these have improved the processes and health interventions in some localities, to determine whether health-related mobile applications have a certain degree of acceptance among the population , acceptance is a central point in the evaluation of medical applications because it determines whether users found affinity towards the given applications, in the present study the usability of both the chat Bot used and the web platform that supports it is measured as a way to determine the ease of learning, the effectiveness of use and the satisfaction perceived by the user.

Continuing with the fields of opportunity of mobile applications, an important development niche would be the one generated by the increase of chronic diseases in the population, which has become a serious problem for health, this is where mobile applications focused on health intervene with their part of support, offering a new perspective on health care , some studies investigate the effectiveness of this type of applications associated with self-care of patients with chronic diseases (Rey Iborra, 2019)

There are also important studies in which the usability of mobile applications related to health is captured, this feature has to do with the ease of reading, fast downloading of information and simplicity in the general handling (Cruz Zapata, 2018) , this feature mentioned above, is shared among some of the applications mentioned in this study with no other purpose than to obtain complete satisfaction in users, some applications leave aside this part focusing on functionality.

The dental area is not exempt from the interest in the development of mobile applications at its service, and proof of this is that there is research that provides tools to improve their development, execution and management of clinical trials in dental health personnel (Bacilio Ruiz, 2021) , as in the application used in this study, the participation of health professionals is vital due to the contribution they can offer in the development of the applications, enriching their content and aligning them with the purpose for which their development was proposed.

Some developments are more directly related to the objective of the present research work and deal with managing treatments against drug addiction and alcoholism, serious addictions and considered a health problem, (Pacheco Campoverde and Idrovo Tapia, 2014) , others deal directly from prevention with the problem of addictions, providing useful tools to identify risk factors and consumption behaviors (Alemán Cortes, 2017)

Other studies linked more directly to desktop applications but with extensions for mobile devices, work hand in hand with augmented reality, based on stimuli and projections, generate environments that, according to the authors help in the treatment of certain addictions such smoking (Pericot Valverde, 2016) , as in the present research, a platform or website that manages and supports the data generated by the chat Bot or the mobile application is of utmost importance to manage and monitor medical treatments and patient records.

In a study related to mobile applications, Sekulovski, (2014), in his thesis, argued that although people still use their cell phones to make calls, the vast majority use them to perform countless tasks through mobile applications, which are available by the millions in the most popular app stores such as App Store and Google Play, providing the possibility to learn, communicate, create, entertain and share information as never before, according to his study in 2014, 22% of the world's population owned a smartphone.

That same year, an exponential growth related to the acquisition of these devices was already foreseen, generating with it an infinity of versions of operating systems, a multitude of screen sizes and resolutions, which generated a large number of consistency problems between the versions of mobile applications, so it is advisable as in this study, to use multiplatform technological tools to avoid these problems of compatibility and consistency, breaking this barrier is intended to a greater number of people avoiding that with them could be excluded from any study that requires the use of mobile devices.

To attack these problems, Sekulovski, (2014), set as the main objective of his thesis, to automate cross-platform application development processes, to achieve this he used a method called Interaction Flow Modeling Language (IFML), this language is used to create graphical and textual models that are used to generate semi-automatically prototyping making development more flexible and agile, thereby reducing the early stages of the project.

In the end, he concludes that cell phones are here to stay, as are mobile applications and that cross-platform mobile development helps programmers create applications faster, with fewer resources, reusable and provides them with assistance to increase their market reach (Sekulovski, 2014) .

As already observed, the increasing consumption of mobile applications is inevitable and its growth goes hand in hand with the number of users that use them, therefore, the strategies for their development must also evolve with the passage of time to meet this great demand, although in the way of their development they may encounter difficulties, for example, fragmentation, which does not allow sharing the same application between different scenarios, fragmentation slows down the possibility of sharing the application without making adaptations in the other environments, which generates a large amount of work and time consumption to generate different versions of the same application for various types of devices (Vique, 2019) .

In a study linked to the use of mobile applications focused on health by Puerto et al, (2016) in which the objectives were "to identify the use and acceptance of mobile applications (apps) in health, in adults attending outpatient Internal Medicine in a regional hospital, through a telephone interview in a sample of 452 patients" (p. 271), a study conducted in the country of Colombia, was related to the study conducted in this document in terms of the technological tool used and the medical area where it was implemented as well as the way in which patients were approached through surveys.

The authors conducted a descriptive study through telephone surveys in a simple random sample, a total of 452 patients were surveyed, a sample obtained according to the total number of patients seen in 2014. When analyzing the results, they concluded that the higher the level of education, the higher the proportion of smartphone users, They also determined that age is a determining factor for the use of these devices, because there is a tendency of lower use at older age, and also, something that was also decisive was that the locality where the study was conducted has a use of mobile devices below the national average of that country.

On the other hand, they also found that the use of mobile health applications in patients attending regular care consultations is a new issue and from which they obtained that only 2.4% of respondents use mobile health applications, and they add that the low percentage is mainly due to the limitations of internet access is that locality and the lack of knowledge by people of the use and management of this type of applications

Ruiz et al. (2015) presented an important literature review showing studies on "mHealth", a term adopted by the authors of the project to define applications that present an enormous contribution to improve access and quality of health services in the Latin American country of Peru. Ruiz et al. (2015), consider that "because health services in developing countries have limitations and are not equally accessible to people living in urban and rural areas, mobile technologies emerge as an innovative option for healthcare" (p. 364).

The study consisted of conducting bibliographic searches of articles on the use of mHealth in Peru using different search strategies in different digital libraries. At the end of the searches, a total of 246 publications were found, of which the majority were excluded because they were not directly related to the topic, leaving only 24 texts to be evaluated in full text.

At the end of the review it was concluded that technologies that include mobile devices are, in general, well received by the population and that their appropriate use in the health sector would help to reduce the limitations in medical assistance. Another conclusion reached by the authors was that the distribution of health-related information by means of mobile devices, remote data recording and remote diagnosis increase the effectiveness of government programs associated with public health, reducing the costs of medical care.

Among the limitations found in the study, it was determined that most of the studies included in the study were conducted in the cities of Lima and Callao, so the results are not representative of the Peruvian community and its implementation in remote and vulnerable places require further studies to prove its effectiveness (Ruiz et al., 2015) .

Finally, they conclude that the use of mobile applications in health is low, specifically in the year 2014 when the information was collected through telephone calls and specifically in the group of patients attending regular consultation of Internal Medicine of the Regional Hospital of Duitama in the country of Colombia.

In the same year, but in the United States Vonholtz et al., (2015), set up an exploratory study on which smartphone health apps patients use, how they are used and how information is shared in health apps.

We surveyed various patients seeking emergency medical care an urban health center, respondents answered questions related to mobile devices on: use, knowledge and characteristics of health applications, for the sample included stable adult patients during the period of the month of April 2013 until September 2013, for data analysis descriptive statistics were used to characterize the demographic sample, the use of health-related applications installed by study participants, of 517 patients, of which 452 met the eligibility criteria and 300 completed the total surveys.

Of the 300 participants, 70% owned a cell phone and 40% used various types of health applications. In the end, they concluded that although mobile applications have experienced enormous growth in recent years, the use of health applications among the sample analyzed had a downward trend, the most used health apps covered the topics of exercise, diet and riddles, they also found that participants most frequently shared information about health apps within their social networks, recommendations by physicians on the use of health-related mobile apps had little response from patients (VonHoltz et al., 2015) .

The results and conclusions presented in the previous paragraph differ from some others presented in this document because contrary to what the authors observed, mobile applications related to health have a high degree of acceptance within the implementation of their research and these reflect that this type of tools are increasing and reflect good results within their studies.

As an example, in Mexico, specifically in the city of Zacatecas, an application for mobile devices called Diabetest was proposed in 2017 by Velázquez-Macias et al, the creators referred that said application supports in the prevention of type 2 diabetes in people over 18 years old, for its creation methodologies were used for the development of mobile applications and a client server architecture, in which all requests are processed either for registration or consultation of information made by customers, the application is multiplatform and can be installed on Android operating systems, or IOS even the web version available on their site can be accessed.

As a result, a stable application was obtained that has the capacity to provide healthy eating and exercise tips, as well as to issue a diagnosis to determine the degree of incidence of the disease using the criteria established by a European diabetes risk assessment scale developed by the Finnish Diabetes Association, this instrument is called FINDRISK (The Finnish Diabetes Risk Score).

The authors concluded that, it is advised that after using this application users should know the diagnosis issued by a medical specialist about their current health and that for no reason this application replaces the diagnosis issued by a health professional, they also added that its use will raise awareness about eating habits and physical exercise, actions that are essential to prevent and/or delay the symptoms of the disease (Velázquez-Macias et al., 2017) .

Subsequently, in 2020, a study was published by Velázquez-Macias et al., as a continuation of the previous work and with the objective of determining the risk of type 2 diabetes in the municipality of Fresnillo, Zacatecas, they tested the Diabetes application for mobile devices, collecting statistical data from this population.

As part of the methodology, a diagnostic test validation study was conducted in 2019, selecting random participants over 18 years of age from the municipality of Fresnillo, Zacatecas, to use all the functions provided by the Diabetest mobile application, with a total sample of 384 participants, which guaranteed the representativeness of the city in the study by means of statistical calculations.

As for the results, it was determined that 25.78% (99 individuals) of the female sex and 21.35% (82 individuals) of the male sex have a LOW risk level of contracting the disease, while 28.9% (111 individuals) of women and 23.95% (92 individuals) of men have a considerable risk level of contracting the disease, while 28.9% (111 individuals) of women and 23.95% (92 individuals) of men have a considerable risk level of contracting the disease.95% (92 individuals) of men presented some type of considerable risk, located between "Slightly elevated" and "Very High", in relation to the possible acquisition of the disease, and it was also observed that the female sex presented a greater risk of contracting the disease.

As conclusions the authors mentioned that; there is a statistically significant relationship between the value of the body mass index of each person with the risk of contracting diabetes, the higher the index the higher the associated risk (Velázquez-Macias et al., 2020) , they also observed that people quickly adopt the use of the mobile application, the management of the application was smooth, this allowed to concentrate the information quickly and automatically to process and analyze it.

As a challenge to be achieved in their study, Velazquez-Macias et al. (2020), state that;

> The application is accepted and disseminated by official health authorities among the population to generate greater awareness of this disease, and in turn to collect statistical data on eating habits to determine actions and campaigns to help contain or further reduce the risk of contracting the disease. (p. 50).

In another research conducted and published in Brazil, which also relies on the use of an application for mobile devices, and which is based on the results obtained in a doctoral thesis, Mazera and Gonzalez, (2018), expose issues related to the measures of protection and assurance of the social rights of people suffering from chronic renal failure and who are in the waiting queue for a transplant, all through a technological platform that also shares information on associated treatments, the main objective of their work was to expose the benefits obtained with the implementation of the mobile application "Kiga".

As part of the methodology used, qualitative interviews were conducted with the patients involved and their families, professional teams, and the data obtained gave rise to the following lines of information: information on chronic kidney disease and kidney transplantation, information on kidney transplant surgeries and, lastly and most importantly, the development of the mobile application called "Kiga".

As results, the authors obtained that there are gaps in care, which do not allow the effective implementation of public health policies. The objective of the mobile application as such was to reduce misinformation about the disease and treatment in addition to offering and accompaniment to patients in the self-management of their treatment (Mazera and Gonzalez, 2018) .

One more study, developed by Bonet et al., (2017), was focused as a literature review and with the objective of "conducting a systematic review of the literature, which allows to obtain an overview of the state of research in the field of interventions with mobile applications in patients with psychosis for the improvement of adherence to treatment." (p. 169), in this work the authors adopted the term e-Health to refer to e-health technologies which combine the use of electronic communication and information and communication technologies (ICT), with clinical, ethical, educational and administrative management uses, with the aim of optimizing health systems in every sense.

In this study they used recommendations and parameters from the PRISMA statement, which provides guidance for publishing research by improving the integrity of reports of systematic reviews and meta-analyses (Hutton et al., 2016) , since 2009 various authors and researchers have used this statement to plan, prepare and publish their work

Following this technique Bonet et al., (2017), selected studies focused on the analysis of the acceptability, feasibility, use and possibilities of therapeutic intervention using mobile applications for the management of patients with psychosis, excluded sources where only phone calls were used, those that were outside the period from 1990 to 2016 and in addition to all those that were not in the English language.

As a result, a total of 431 articles were identified and reduced to 112 after eliminating publications not aimed at patients with psychotic disorders. A further 92 publications were subsequently discarded because they were not of the clinical intervention type but rather systematic reviews, surveys and study protocols, to finally reach a sample of 20 articles, of which 17 were independent interventions.

In conclusion, the authors consider that the mobile interventions of the analyzed works show a feasible maneuver for patients with psychosis, also when studying the responses of those involved, most are satisfied with these interventions, they find them useful, beneficial and easy to use, all this suggests that these interventions are appropriate and well accepted by patients, also note that although the results are generally positive, there is a minority that has expressed difficulty in the use of mobile devices and consider a huge number of communications per day which translates into something intrusive and tedious.

They further state that interventions for other affectations show some kind of benefit, in this sense, although the results were not conclusive, they did manifest the wide use that can be made of this type of technology.

Studies related more specifically against smoking, indicate the level of severity caused by this disease and the consequent damage it causes, such is the case of the report of the World Health Organization called "WHO Global report on trends in prevalence of tobacco smoking" or in Spanish "WHO Global report on trends in prevalence of tobacco smoking" published in 2015, determined that tobacco is the only legal drug that kills many of its consumers when used exactly as determined by cigarette manufacturers, WHO has estimated that tobacco use is currently responsible for the death of approximately six million people worldwide per year and about 600,000 people who are also estimated to die from the effects of secondhand smoke (WHO, 2015) .

Therefore, it is of vital importance to support the problem of smoking, through the use of information technologies, specifically Bots, which represents the main objective of this research work, and others that are mentioned in this document, it is also essential to evaluate their performance and interpret their results for further improvements.

There are several research works that make their contribution to this serious global health problem by implementing the use of technology in support of anti-smoking treatments, as an approach to this type of work, León et al., (2015), conducted a review project of scientific studies that according to their main objective;

> It presents a review of the scientific studies that have been developed in the process of smoking cessation, which allowed to support a research project that will adapt a mobile technology model designed and used in the United States in English-speaking and Spanish-speaking populations (p. 1999).

They also analyze the benefits of incorporating mobile technology as an effective strategy in the treatment of smoking in the process of smoking cessation. The methodology of this work was carried out in two phases; the heuristic phase involved the search for scientific papers and the second, the hermeneutic phase, was based on the analysis of the selected texts categorized by predefined sets.

The results they obtained mention 14 scientific articles related to the use of mobile technology as a support for smoking cessation; after an exhaustive filtering, the number of works was reduced to five studies, among them it is highlighted that their main objective was to measure the effectiveness of an intervention based on this model based on the use of mobile technologies. In the end they concluded that cell phones can be an effective tool with potential to develop programs aimed at promoting and addressing health problems related to smoking (León et al., 2015) .

In a similar study, funded by the Ministry of Science and Innovation of the Government of Spain, and with the intention of obtaining a PhD degree, Pericot Valverde (2016), performs a work on virtual reality techniques in the treatment of smoking, with the aim of developing and empirically validating a procedure through virtual reality in the treatment to mitigate tobacco cravings.

As conclusions, they summarize, among other things, that virtual reality as an exposure technique is a viable method to simulate daily life situations associated with tobacco use and that the technique of exposure to clues through virtual reality has the ability to reduce the anxiety conditions experienced by smokers.

García-Pazo et al. (2020) made another literary review contribution to the field of the use of technologies against smoking and considered the use of mobile applications to quit smoking, arguing in their research that smoking represents a health problem that is difficult to suppress.

The most addicted and dependent people to nicotine present other problems such as depression and anxiety. The aim of their work is to conduct a literature review of mobile applications for smoking cessation that apply Cognitive Behavioral Therapy and describe the techniques they implemented. Using the PRISMA methodology for conducting literature reviews (Hutton et al., 2016) , same methodology also employed in other works mentioned in this paper

The authors conducted a search during the period from 2010 to 2019 in various databases, finding a total of 415 papers of which only five articles were the subject of research, the rest were excluded after applying some criteria determined by the authors. The applications studied within the articles contained various functions, among them; consumption records, visualization of progress graphs, educational videos and motivational sections.

The results showed the need to include in this type of application some analysis of the smoker's behaviors, since some of them do not have them, and a direct line of communication within the application between users and the health personnel in charge is also necessary so that they can provide the most appropriate treatment.

As a conclusion, they highlight that the systematic review carried out has highlighted the lack of mobile applications that support actions to quit smoking, the applications detected have some limitations by not providing sufficient information regarding the techniques used in Cognitive Behavioral Therapy, they only report some procedures they use, but without further details, if they did it would facilitate the standardization of the program used (García-Pazo et al., 2020) .

Recently there have been found works directly related to the topic raised in the present thesis making use of Bots as a main support tool for health issues, within these recent studies is the following thesis work published by Labra Chino and Quispe Poma (2022), in the Latin American country of Peru, and which consisted according to its authors in;

> To propose a reference method for the attention of health consultations based on chatbot, avoiding face-to-face consultations; for which a reference method was elaborated to design and develop the chatbot, with the purpose of answering users' queries online and immediately (p. 34).

The study was conducted in a health care center for older adults, and the Bot was designed to answer specific questions about COVID-19 based on content published on World Health Organization (WHO) web pages.

As part of the methodology implemented by the authors, they highlight the implementation of 4 phases that determine the use cases, the scripts, the design and finally the development, as results they argue that, based on the metrics used, the chat Bot is efficient in terms of response times, it is flexible and optimal in terms of functionality, a star rating system was also used to determine the overall acceptance by users.

In the end they concluded that the chat Bot contributes to have a significant improvement in the process of attention to consultations, in addition to this the authors determined that the chat Bot prevented people from going in person to the health center and that this practice drives the digital transformation and with it a competitive advantage is obtained, in addition, they assured that the area of research related to the use of chat Bots in medical issues is little studied and there is not much information about it (Labra Chino and Quispe Poma, 2022) , this study coincides in a certain way in what was argued in the conclusions by the authors to what was exposed in the present work regarding the amount of information related to the use of Bots in medical issues.

The topic of human health has not been the only one benefited with this type of projects, there are chat Bots involved in animal health issues, which facilitate interaction with veterinarians to people who own pets, one of these contributions refers to the use of this type of technology, specifically the thesis work of Rodriguez (2022), One of these contributions refers to the use of this type of technology, specifically the thesis work of Rodriguez (2022), conducted in the city of Queretaro in Mexico, which shows the use of a chat Bot as a tool to interact by solving doubts to people who own dogs, either questions related to emergencies, behaviors, or general care, the tool works based on their geographical location to put the user in contact with specialists in the area.

In their methodology they implemented concepts associated with artificial intelligence so that from a knowledge base the system responds according to rules and grammatical algorithms with the best response for the user, also emphasize that this type of applications do not replace the opinion of an expert in situations of risk or special care. As a result, they obtained a prototype of a functional conversational agent or Bot, implemented on a web platform.

In the end they concluded that the software is a system that is growing and that its users are the ones who provide the information for the Bot to acquire new knowledge through the related questions, because like all artificial intelligence, it is constantly learning, the questions that are not in its knowledge base cannot be answered (Rodriguez, 2022) , contrasting the conclusions with those of this and other similar studies, the Bots are limited to the tasks for which they were designed and it is desired that they meet new specifications will have to be programmed to expand their knowledge base.

Another contribution is the case of the following work with the publication format "letter to the editor", we found the analysis carried out by Segrelles-Calvo et al, (2021), in which they conducted a literary search of publications related to the use of chat Bots in medicine or specifically in the topic that concerns us, the fight against smoking, the works are similar to those presented in previous paragraphs of this document in both objectives and results, but we can highlight the conclusions reached by the authors related to the limitations regarding the use of chat Bots.

The authors mention that there is a limitation at the time of human interaction because it is not possible to have a personalized treatment which could lead to some kind of damage in patients if not detected in time, these interactions should be fully endorsed by medical professionals, on the other hand they mention that this type of technology has great potential to combat smoking but before it would have to demonstrate its effectiveness, in the end they recognize that there is currently little evidence but this, although minimal is hopeful, (Segrelles-Calvo et al...), 2021) .

The field of application, nowadays for chat Bots is very wide, as a sample we have the contribution of the thesis of Diaz Guerra (2021), in which he exposes the work of a chat Bot called OncoBot for learning the prevention of breast cancer, where he addresses issues related to this type of cancer supporting the prevention of early diagnosis of breast cancer, whose main concern is the high mortality rate as a result of late diagnosis, as the main objective of his work was to determine the effect of increasing knowledge using chat Bot for learning the prevention of breast cancer.

In his methodology the author used a type of experimental applied research, a pre-experimental design followed by the treatment and finally the post-stimulation test, in addition he used the pre-trial design because the previous trial was conducted in group 1 without the use of the proposed tool.

As results the author obtained that, in relation to the knowledge of the 34 people about breast cancer prevention, the percentage of knowledge increased to 70%, the percentage of motivation increased to 76% and the percentage of satisfaction increased to 81%. In conclusion, the author determined that according to the results obtained, the learning related to breast cancer prevention through the chat Bot called OncoBot, has a positive effect on the group of people who participated in the study, this observed in the increase of knowledge, motivation and increased satisfaction that was achieved after the interactions by the participants who have been using the chat Bot OncoBot (Diaz Guerra, 2021) .

Another thesis that deals with the same line of research in this regard is the one published by the authors Cruz Barrera and Zambrano Lazarte (2020), who claim that the tool is able to "help users to reflect and reach a higher level of knowledge about sexuality" (p. 9), the main objective set by the authors was to determine the effect of a chat Bot for learning about sexuality, as a methodology an applied study and a pre-experimental design was implemented, as for the software development they used a SCRUM type methodology, and to conduct the study a sample of 60 people in the city of San Juan de Lurigancho.

As results the authors determined that the use of the chat Bot increases the level of knowledge by 33%, motivation by 84% and user satisfaction by 80%, at the end they concluded that during the development of the research it was possible to determine the effect of the chat Bot for learning sexuality as successful since it met the objective by increasing knowledge, motivation and satisfaction (Cruz Barrera and Zambrano Lazarte, 2020) .

Preceding the research described in previous paragraphs, Ávila-Tomás et al, (2020), also conducted a study involving a chat Bot called Dejal@Bot, and the treatment against smoking, in which, although they do not highlight in their publication technical aspects such as programming languages, operating systems, internet support platforms used, libraries, messages, among others, neither do they mention in the study a clear methodology of the monitoring carried out for the development of the Bot, what they do mention are some of the guidelines that must be met by this type of applications.

These guidelines, supported by professionals are that their contents are based on scientific evidence, that they offer security and privacy, and that users require a customizable, low-cost tool that helps to deal with withdrawal symptoms and side effects.

At the end they conclude with the following statements; there is a primary need to evaluate digital tools after being implemented to demonstrate their validity, common languages must be generated in an environment that transcends health professionals is a current need, this with the intention of having effective communication between health professionals and with computer scientists and software programmers also as a last point should be taken into account the assessments and opinions of users and professionals before starting any process of developing any application (Avila-Tomas et al., 2020) .

A few years ago, in 2017, Velázquez-Macias, et al. published an article from which the present research work entitled "Desarrollo de un Bot para apoyo en el tratamiento del tabaquismo en el Centro de Integración Juvenil en Zacatecas" was derived, they developed a Bot as a support in the treatment against smoking.

In this article, they graphically expressed the use of the Bot in collecting smoking habits among patients, replacing physical means of registration, the Bot's functions were limited to storing information provided by patients and therapists and providing tips with health advice, the Bot did not have the ability to diagnose or determine individualized treatments.

As a conclusion, they argued that the development of Bots is a feature of the software little explored in this type of interventions, by developing the project gave support and support, by serving as a complement in the treatments of patients with smoking, that health professionals indicate. (Velázquez Macías et al., 2017) .

In addition to the Bots focused on the treatment against smoking there are also others that support other medical areas, as expressed by Pérez Peña and Ramos Jurado, (2021), in their thesis entitled "Chatbot with artificial intelligence for the process of customer service in the Urology Service of a health facility", the authors detail the benefits of the chat Bot;

> Improving the processes of patient care, allowing the optimization of the resources involved in the operation, through the use of the web, the flow of patient care was analyzed and improved, and an alternative was presented for making appointments and creating payment orders for the different services provided. (p. 8.

In addition, they describe that the tool improved the level of satisfaction of patients and attendees at the health center during peak hours, this being the main problem in the study.

The general objective of his thesis was to determine how a chat Bot synonymous simply with Bot or more technically defined as autonomous personal assistant, which had some principles of artificial intelligence, improves patient care in the urology service of a health center, through the use of a web application was examined and improved the flow by patients, an alternative to the management of appointments and the process of charging for the different services offered was proposed, thereby improving the percentage of satisfaction of patients and visitors to the health center in the range of peak hours, recognizing in these the main problems that arose in the study of the study.

As a methodological process for his thesis, the survey was used as a technique to evaluate the study variables, which were those assigned to urology patients within the care process. The instrument used for data collection was a questionnaire composed of 40 items, the reliability of which was determined by Cronbach's alpha coefficient with the support of 30 patients who used the Bot and answered the questionnaire confidentially and objectively according to their personal perception.

At the conclusion of the research it was established that an adequate use of Chat Bot by patients, significantly improved customer satisfaction, according to the authors, by 78.92% determined as an acceptable result. (Pérez Peña and Ramos Jurado, 2021) .

In another more specific study related to young people and the issue of sexual health conducted in a school in Chile, Barker Maillard (2019), aimed to evaluate and develop a Bot with a virtual assistance system on sexuality based on Artificial Intelligence tools in a school in Chile.

The Bot was designed to be used as a complementary tool to strengthen sexual education topics in middle school students, by collecting statistical survey results through a quantitative data analysis plan.

From the observation of the responses and the focus on the main research questions of their study, they determined the following results; the perception by the users of the Bot called Isidora in general terms was positive, due to the fact that 73% of the students who argued to be satisfied with the Bot, while 86% mentioned that they wish to continue with access to the tool, 82% said to have a positive influence with its use and 97% assured that it is a useful instrument for adolescents.

As a conclusion they mention that the Bot has a great potential to function as a support tool in the education of a varied group of adolescents and also answer doubts about sex education helping to strengthen their knowledge in order to have more responsibility about the subject (Barker Maillard, 2019) .

Five years behind the previous study already in 2014 developments in this area were beginning, a research work conducted in Mexico, in which Cabrera Mendoza et al., (2014), designed and developed a mobile messaging system called mSalUV, which allowed patients with type two diabetes mellitus to remember dates and times of their appointments in addition to timely and on-time intake of medications, on the other hand, promoted healthy lifestyles and, in the background gathered statistical data to subsequently evaluate its effectiveness.

The methodology of the work was divided into three phases; the first covered the design and development of the application for mobile devices mSalUV, the second was used to generate and write the text messages to be sent to three different branches and the third to find out the opinion of the users regarding the use of mSalUV, the study included a total of 46 patients previously selected and which met the requirements defined by the authors, in addition, about 40 text messages were designed and approved by the medical authorities.

During the 45 days of the period contemplated to collect data, a total of 1850 messages were sent, users argued that the mSalUV mobile application helped them in the treatment of their disease, which in turn was easy to use and showed interest in continuing to use the application in the future, the conclusions reached by the authors define the system as acceptable by users, people with diabetes mellitus 2, which poses an interesting scenario to take advantage of new technologies for the benefit of health

2.2 Theoretical Basis

2.2.1 e-Health

The term e-health has been used by some authors to describe medical care practices based on electronic processes and communications, the World Health Organization defines it as the cost-effective and safe use of ICTs in support of health and its fields, including medical care and surveillance services, literature, education and research related to health issues. There are other descriptions of the term, for example, León et al. (2015), mentions that the term e-health refers to health care and practice with the help of information technologies, in agreement with Fernández (2020)

The term, apart from referring to information technologies in support of health also mentions that it has to do with medical environments at various levels; management, prevention, diagnosis, monitoring and treatment, Bonet et al., (2017), adds that these technologies are combined with clinical, ethical, administrative and educational concepts, with the aim of strengthening the health system, and facilitating greater access by the population to health.

The study and the associated tool presented in this document fits perfectly into the term e-health in a globalized way because within its methodology technological means are implemented within the administration of its work, whether preventive, informative, care or treatment.

2.2.2 Mobile health

The term mobile health, m-health or m-health is considered part of e-health or electronic health, León et al, (2015), expresses that this concept refers to the component of e-health that uses mobile devices such as cell phones, tablets and pocket electronic devices to provide health services, Ruiz et al ., (2015), adds that these devices also serve to transmit and provide medical assistance and information, but not only entails the use of this type of devices also takes into account the services that are around them such as for example the use of text messages, wireless transmissions, voice calls, and mobile applications to transmit health-related information (Betjeman et al., 2013)

Other contributions to the definition also consider the global environment of the use of a cell phone and the platforms and technologies that support it, such as global positioning systems or GPS and Bluetooth technologies (Rowland et al., 2020) .

The tool associated with the present work for the registration and monitoring of patients in treatment against smoking is considered within the category of those involved in mobile health, the Bot when working of a mobile application is considered as a whole as such, therefore, such application is within the category of those considered as mobile health.

2.2.3 Artificial Intelligence

To try to approach a more or less standardized definition of the term Artificial Intelligence, most authors go back to the year 1950 when Alan Turing proposed a set of criteria to determine whether a machine can be (or not) as intelligent as man (Bistarelli et al., 2012) , based on these criteria some authors have tried to propose some definitions but without agreeing yet on a precise and universal definition, the term as such of artificial intelligence was proposed in 1956 by John McCarthy to refer to the branch of computer science that was dedicated to the study and design of intelligent machines

From there some researchers concluded that the Turing test is enough to define artificial intelligence (Alfonseca, 2014) , some authors go a little further with their definition arguing for example that Artificial Intelligence is the ability of machines to use computer algorithms, learn from data and use it in decision making similar to human action (Rouhiainen, 2018) , some others consider Alan Turing as the father of Artificial Intelligence giving full credit to his claims and conclusions (Estrada Cutimbo, 2018) .

Currently there are two aspects of artificial intelligence, the first one named as General Artificial Intelligence related to the ability to solve tasks, such as thinking and acting similar to the human mind. The second named as Narrow Artificial Intelligence or Weak Artificial Intelligence related to the ability to perform specific and repetitive tasks (Leyva-Vazquez and Smarandache, 2018) .

The Bot by its intrinsic nature fits perfectly in the second aspect because it has very specific capabilities and could be considered repetitive, since everything in its programming is defined to respond to certain previously defined commands, it could be considered a type of artificial intelligence too basic because it does not have the ability to learn new actions or respond to new commands introduced by the user.

On the other hand, the Bot is only able to answer those questions for which it was programmed and actions not recognized can be easily interpreted with default answers or non-recognition messages. This does not prevent its attributes to be used in a practical way or to be limited by the aforementioned functionality, since as time goes by they can increase their knowledge base according to criteria observed by the programmers based on experience.

2.2.4 Personal assistants

When referring to a machine with Artificial Intelligence it can be referred to as a robot when it is composed of hardware components, and Bot when it is composed of software components, both of which are also referred to as agents. Some agents that are currently having a great boom are chat Bots and personal assistants. A chat Bot is a Bot that interacts through messaging by means of text strings, on the contrary an intelligent personal assistant is a Bot that performs certain tasks and offers services of different types such as internet-based searches or activation of compatible peripherals, (Rabelo et al., 2018)

Personal assistants or virtual agents respond to more complex questions and in turn can learn over time the habits and preferences of users, today there are different options on the market such as Alexa, Siri, Cortana, among others.

These commercial options offer different functionalities, such as voice recognition, service subscriptions, integration with smart peripherals, among others. These devices have had a significant boom among consumers, according to a study conducted by Microsoft in 2019 called "2019 Voice Report: consumer adoption of voice technology and digital assistants" mentions that new voice recognition technologies have given to a new generation of digital and voice assistants

Consumers can now interact with search engines on a deeper level and in a more meaningful way by the power of their voice and with the support of artificial intelligence, further mentioning that in their study 72% of respondents reported using a digital assistant in the last 6 months, this report discusses the latest voice recognition technology and the artificial intelligence technologies that support them

The results are based on two consumer-focused surveys and internal Microsoft data on a variety of topics including; adoption and use of voice technology in digital assistants, consumer confidence in voice technologies. digital assistant functionality and voice recognition skills, and the evolution of e-commerce (Olson and Kemery, 2019)

Although there are several options in terms of devices that function as digital assistants, some of them are little known among consumers, the large monopolies have completely dominated the market, proof of this is that only 1% of the market is represented by these alternative optionsFigure 3 shows the distribution of users worldwide of the most popular personal assistants in 2019.

Figure3 *Percentage of personal assistants usage in the market in 2019.*

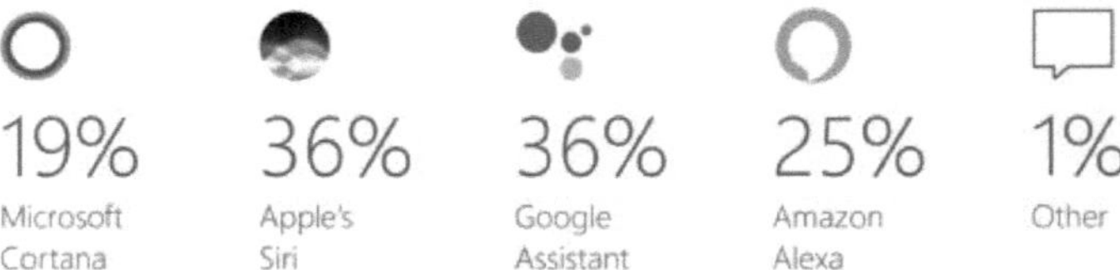

Note. taken from (Olson and Kemery, 2019, p.x. 9).

2.2.5 The Bots

Bots or chat Bots are considered conversational agents which respond to scripts already predetermined in their programming phase, they can be defined as robots that interact with people through chat messages, pretending to be an operator or a person in real time (Leyva-Vazquez and Smarandache, 2018) , the communication of a Bot is very simple, it only responds to questions asked by the user (Dahiya, 2017) and its answers according to Guerrero et al., (2017), are based on knowledge bases and represented by natural language text messages.

In other words, a Bot is an online human-computer dialogue system by means of a language known to the human, the Bot is considered an agent embodied in technology by the fact of providing a sense of presence (Cahn, 2017) , even as an optional way some were given names in order to generate trust in the user and to fulfill the previous question, for example, Selena (Velázquez Macías et al., 2016) , or Geraldine (Velázquez Macías et al., 2017) .

A Bot furthermore, is a software application that is added to a messaging application solution as a contact and offers through interaction with a web service and a database a response returned also as a text message , These programs are designed to simulate how a human would behave as a conversation partner, but with no intelligence of its own, except that given to it by its programmers (Barker Maillard, 2019) .

Currently there are countless Bot programmed to meet the needs of different disciplines, some more complex than others, but all of them fulfill the basic function of interacting with human beings through text messages, among which stand out Bots for help on Web platforms, Bots for weather and air quality consultation, Bots focused on health issues that are of interest in this research work, some more specific Bots to perform certain tasks which are not available to the general public and are reserved for employees or managers of some large companies. Some Bots have had enough acceptance among companies dedicated to e-commerce and consumer users, these provide a change in the shopping experience and customer service by companies towards their potential buyers, a bidirectional interaction between the consumer and the business intervenes, they have great advantages because they offer instant services at any time of the day, provide help through a bank of questions, also allow simple purchases, provide offers and news and can send alerts about available inventory (Chesñevar and Estevez, 2018) .

2.2.6 Applications for creating Bots

Nowadays there are countless mobile instant messaging applications such as WhatsApp, Line, Snapchat, Facebook Messenger and Telegram that share most of their main functions. Telegram Messenger is an application developed and maintained since 2013 by brothers Nikolai and Pavel Durov.

One of its main functions is focused on sending and receiving text and multimedia messages, it is managed by a non-profit organization with operations center in Dubai, United Arab Emirates (Galán Pache, 2014) . It supports the hosting of all types of files, one of its main capabilities is to support the Bots platform, which allow the creation of interactive conversations through predefined dialogues (Benito Rodríguez, 2018) .

Telegram is the only instant messaging application in 2020, which provides the ability for its users to create Bots for a variety of applications according to their needs, (Mulyanto, 2020) , even, users can develop Bots to manage payments, develop games, moderate groups, automate tasks, among others, all thanks to the principles of artificial intelligence. It is compatible with most Android-based systems from Google and IOS systems from Apple and is also compatible for logging into modern browsers (Gil and Afrashtehfar, 2020) .

For the programming of the Bot, the Python programming language created by Guido van Rossum in the early 1990s was used, whose name is inspired by the English comedy group "Monty Python" (González Duque, 2011) . Also implemented with the libraries provided by Telepot, which help to create Bot-type applications using the services provided by Telegram, (Rodrigues et al., n.d.) .

At present (2022), most instant messaging services provide the ability to generate Bots as part of the services they provide to users, thanks to the release of their libraries, some of them with commercial support for those who can afford a license and some others free for any user to generate, develop and support the specific tasks required through a Bot.

2.3 General project-related concepts

Throughout the document various terminologies and idioms will be used, some of them Anglo-Saxon, which were adopted in the Spanish language and are not translated literally; these technical terms are directly related to the project and serve to contextualize the complex organization of the entire project;

2.3.1 The software

According to Sommerville (2010), software is not only computer programs, but also involves the associated documentation and data configurations that make these programs work the way they were created for , these programs together form more complex applications which can be desktop applications, webs, even mobile, among others, Sommerville (2010).

To complement the definition, the actors that perform these tasks, in this case the professional programmers, are the ones that maintain, update and give maintenance to the programs for a certain period of time (Pressman, 2010) , in addition to this, the management of another concept is required, the "software engineering" to produce software with quality based on methods, principles established and validated by the same.

Since Bot is written in a programming language, in this case Python, it is considered an applicative software and meets the basic conditions to be considered a software, since it needs updates, adjustments and error correction, it also has the characteristic of being intangible and its storage is done through digital media, such as mechanical or solid state hard disks or in a service provided in the cloud.

2.3.2 Software engineering

Sommerville, (2005), states that the term software engineering is derived from a branch of engineering, which is the main support for the development of software applications, in which systematic, organized, disciplined and quantifiable guidelines are applied to facilitate the development to create high quality software.

Software engineering focuses on the problem of software production, its managers, software engineers require the necessary skills in computer science . While Pressman (2010), believes that processes, methods and tools enable software engineers to fulfill their task, software development.

All software development should be aligned to a methodology recognized by software engineering to meet the requirements of quality, functionality and compatibilities needed, also to have a solid documentation control that allows updating depending on the needs of the customer and users and correct any errors generated from changes in software or erroneous inputs generated by users.

A chat Bot combines all the complexity that allows software engineering with all the complexity of natural language processing, this provides the necessary solids so that they can be implemented on different platforms often will have to use third-party services or manipulate databases from different vendors, they may even activate remote sensors or smart peripherals compatible.

2.3.3 Software Development Cycle

When working on the construction of a product or system, it is important to execute a series of previously defined steps to obtain in time and form the desired product, this set of steps is known as "software process" (Pressman, 2010) . To organize these activities, aspects such as sequences, times and resources must be considered, for which there are different ways of representing the flow of tasks, which must be considered according to the type and size of the project

In the development of the Bot involved in this document, the linear software process method was selected, for which Pressman (2010), considers five structural activities for the development of the general actions in software engineering, which are ordered hierarchically and work one after the other as shown below; communication, planning, modeling, construction and deployment.

These activities can be ordered depending on the needs of the project, but in general each one describes the way in which the process flow is developedfor practicality and because the requirements prior to the development of the Bot are clearly known, the waterfall model or better known as the classic life cycle was selected.

2.3.4 Cascade cycle

In software engineering and in general in the field of desktop, web or mobile systems development, the waterfall development cycle, also known as sequential, is one of the different methodologies used in software development.

This particular methodology orders the different stages of the development of the application in a sequential manner where each stage must wait for the end of the stage that precedes it before starting to avoid false steps, at the end of all stages a complete exploration is carried out, if there are any errors or failures in the system, the processes of the phase or phases involved must be repeated in order to correct the code that caused the problem

This development cycle is easily adapted when working on mobile application developments due to the ease of implementation and the logic it follows in its processes which can be easily understood by professional programmers

Pressman (2010), believes that although this system has several advantages in its implementation, some of its disadvantages are the natural risks that every project presents and which alter the sequential flow of activities, another disadvantage would be that often customers do not express the requirements completely at the beginning of the project, and that there will not be a functional version until the end of the process, which often causes customers uncertainty about the final results.

2.3.5 Mobile Application Development Methodology

If you adapt the waterfall development methodology to a mobile application project you will have fast results in a short time, this methodology was the one used in the development of Chat Bot (also considered as an application for mobile devices), it is based on techniques used and accepted in the mobile application development industry, educational software engineering and in the concept of agile methodology for its versatility and fast performance, this methodology is divided into the following stages; Analysis, Design, Development, Functional Testing and Delivery

As can be seen in Figure 4 all the stages are represented, including some of their associated tasks aimed at working and fulfilling the objective proposed in each stage, which will be described in subsequent paragraphs.

Figure4 *Stages of the methodology for the development of mobile applications.*

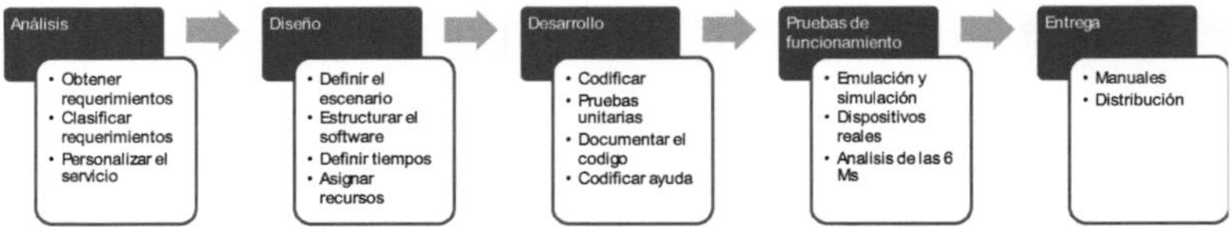

Note. Adapted from Stages of the methodology for mobile application development, by Mantilla et al., (2014).

In the analysis phase, as its name suggests, the needs of the end users of the software or application are analyzed to determine the objectives to be met, the procedures and processes to be carried out, the handling and treatment to be given to the information, it is also considered the most important part because this is the beginning of a good planning and from here the other phases of the project are developed

In this sense the requirements gathering process concentrates on the software, additionally the space and limitations of the software, its functions, possible performance and its interfaces must be understood (Maida and Pacienzia, 2015) , in other words the analysis defines "what" of the software, i.e. what function the new program will perform (Adenowo and Adenowo, 2013) .

In the design phase the activities are divided and ordered in such a way that the components can be developed separately one after the other, i.e. they are modularized, therefore these activities can be distributed among the human resource more easily, for this project two types of design were used; the high-level and the architectural, the first basically defines the logic that the project must follow to obtain and process the information while the second defines the operations and algorithms that software must use to perform the tasks entrusted , this phase describes the "how" of a software system (Adenowo and Adenowo, 2013) .

In the development stage the source code is implemented according to the programming language or languages that were selected by the development area, initially generating software prototypes performing basic tests and validations. The programming language will be subject to the capabilities of the computer equipment and the needs required in the first phase, the libraries and reusable code will be very useful in the more advanced phases of the project streamlining the results with a more efficient programming.

Functional tests are performed on the previously coded modules, they are connected to each other to check the integrity of the information, hoping thereby to obtain the expected results with the correct processing of the information, all this work is a step prior to delivery to the end user of the system

In the last phase of the project and of the waterfall methodology, the end user performs tests to verify that the system performs what was intended at the beginning and that it in turn meets their expectations

2.3.6 Supporting operating systems

On the client side, the Bot uses the Telegram platform to run, therefore, you must have a device that supports this mobile application, fortunately Telegram can be installed on most Android based operating systems owned by Google and IOS owned by Apple, it also has support to run on any modern browser (Gil and Afrashtehfar, 2020) .

An operating system is a type of complex software that serves as an intermediary between applications and all the hardware devices of a computer, tablet, cell phone or any other smart device, it is also categorized as a set of programs that allows to manage the RAM memory, hard disk, storage media and all the peripherals that the device in question has.

The operating system is also responsible for executing processes and programs necessary for the device to interact correctly with the user and respond to user actions in a fast and error-free manner, always taking care of the integrity of the information.

On the server side, the Bot will be in constant and indefinite execution and will need a stable operating system with sufficient capacity to be able to respond to all the requests that at a certain time are usually numerous, in addition to having a correct error management to keep the Bot running as long as possible, this way the users' requests will not be affected and thus they will have a more satisfactory experience.

2.3.7 Programming language

The Bot was developed using the Python programming language created by Guido van Rossum in the early 1990s, whose name is inspired by the English comedy group "Monty Python". It is an interpreted, multiplatform and object-oriented language (González Duque, 2011) . It is also implemented with the libraries provided by Telepot, which help to create Bot-type applications using the services provided by Telegram (Rodrigues et al., n.d.) .

There are other alternatives that present a more advanced level of programming through the so-called frameworks and some other parsing tools which, depending on their capabilities, can be paid, that is, a license must be purchased to use them. On the other hand, the Bot used in this research was developed and implemented in free software both to write its code and to implement it in an operating system.

2.3.8 Databases

It is the primary means for storing structured data, it connects its structures together forming a single module of data grouped into a logical unit, these are used to manage huge sets of information, a database, is thus an organized compilation of information that is originally stored electronically in computer systems (Greenwald et al., 2013) .

In order to have a record of the patients' activities within the Bot, it must have the capacity to store the data for later consultation within the consultation platform designed for this purpose. This process is carried out and managed on the server side, which is a completely transparent process for patients and therapists, following the basic regulations that the handling of a database must comply with: integrity, security, reliability and availability.

2.4 Theoretical approach

The theoretical approach proposed in this document is divided into two parts; the first one refers to the theories that support software development and the second one to the theories that have to do with the evaluation of the tool generated from the standards established in the first part, such evaluation has to do with the measurement of the impact and effectiveness in accordance with the results obtained without the use of the tool and after it

That said, the following methodology refers to the processes involved in the development of mobile applications some of them, for example, are agile methodologies, which have been used since the beginning of the trend towards the development of mobile applications, because they present quick solutions, especially those projects where the requirements are constantly changing.

These applications have to consider a series of characteristics and special conditions, which should contemplate different requirements such as: channel, mobility, portability, specific capabilities of the terminals, developments and projects for mobile devices are usually carried out without an established order and by individual developers who do not implement some kind of methodology related to Software Engineering

This methodology is considered as moderate for its ability to adapt to different environments and as its name indicates the agility in its results, on the other hand there is also the traditional methodology, considered as heavy for documenting all its processes and planning in a strict way which must be well defined from the beginning, impose certain disciplines which make a rigorous work to make the software development process more efficient, these methodologies do not usually adapt correctly to changes, so it is not recommended in scenarios where the requirements cannot be predicted or constantly vary state as time goes by (Maida and Pacienzia, 2015) .

Consequently, and taking into consideration the two alternatives in this project, we declined to use a traditional methodology, since we have all the necessary requirements, both for the design of the Bot and for the web platform that will manage the treatment by the therapists' hands.

Specifically for the determination of the Bot's functionality, and since this is considered within the definition of Artificial Intelligence, it is required a theoretical support that formulates and justifies some of the actions that the Bot in question can perform, all this attached to the already established rules of communication and the determination of the type of responses, for this the theory of automata supports in this sense.

Automata theory is a branch of computer science that studies abstract machines, their response capabilities and, in addition, the problems they are able to solve. This theory is closely related to the theory of formal language and the classification of messages that they can recognize and interpret to subsequently perform a finite action, , actions that the Bot must perform throughout the period of operation.

The theory of formal language is closely linked to the theory of automata because it is in it where languages are defined and how they should be interpreted by machines, (Contreras, 2012) . This theory is also directly related to mathematics, logic and computer science, its language rules define groups of symbols, which later will constitute the vocabulary of the applications themselves, these grammar elements are composed of collections of symbols with their own specific meaning, often only for the environment in which they were designed ,

Continuing with the theories related to software development in this case the Bot, it must take into account the visual impact that the final application will have on them users, because a good selection and distribution of colors, avoids visual fatigue when using the application, here intervenes the theory of colors, described since years ago by some characters such as Johann Wolfgang von Goethe, which proposes specific mixtures that combined correctly produce various sensations that are translated in the human brain such as acceptance or rejection

By using these combinations it is possible to predetermine whether the use of a certain combination will be evaluated with a high or low degree of acceptance, these conditions can be applied to the development of web pages, applications for mobile devices and in general of any software that includes graphical interfaces as part of its components, colorimetry is an important part due to the effects it has on the human brain whether positive or negative (Arias and Vela, 2015) .

The second part of the theoretical approach as already mentioned, derives from the need to measure the results obtained according to the opinion of the end users of the Bot, in this case patients and therapists, by means of some instruments to later interpret their results, this will be detailed later in

Chapter III of this work, continuing with the theme of the tests it is necessary to implement related methodologies with the evaluation of applications for mobile devices

To meet the satisfaction sought by end users when using the application, some tests must be applied to ensure the quality of the software, there are countless tests for software projects which can be of two types; non-functional tests allow to know the risks to which the application is exposed as well as the average performance in execution, and functional tests which focus mainly on the execution by providing feedback on their behavior, also check that they really perform the tasks for which they were designed (Solis et al., 2014) .

There are also more specific tests designed for common software applications that can be adapted with certain modifications to applications for mobile devices, for example;

- Unit testing: this test focuses on the basic functions of the application and how it responds to different scenarios.
- Integration test: this test verifies the integrity of the information as it passes through various modules.
- System test: verifies that each element fits together properly and that the functionality and performance of the total system is achieved. The system test is made up of a series of different tests whose primary purpose is to exercise

- Validation test: based on comparisons and the introduction of different values, this test ensures that the results are as expected after some calculation or internal process.
- Recovery test: a premeditated failure or breakdown is generated in the system to verify its response and recovery capacity.
- Security test: this test verifies the protection mechanisms established in the application's functionality.
- Endurance test: the application is subjected to excessive processes to verify its response, also known as stress test.
- Performance test: the performance of the final product is tested at runtime.
- Installation test: installation is performed on different devices with different configurations to ensure compatibility.
- Regression tests: these are tests that are performed after some minor change and are compared with the results of tests performed before mummification (Maida and Pacienzia, 2015) .

In this context, Pressman (2020), catalogs certain specialized tests designed for unconventional applications, as he calls them, in this case the Bot as it does not fit completely as a conventional application (desktop application), nor as a web application, it is rather similar to a mobile application, but not entirely, instead it is cataloged as unconventional, therefore, the author proposes the following tests:

- Graphical user interface testing. This type of components have become more reused throughout an application and it is common that they present consistency errors when dragging options from a predefined template, therefore, they should be checked to verify that there is concordance with the function to which they refer.

- Testing of client-server architectures: This type of testing is essential in non-conventional applications such as Bots, since a large part of their operation depends on network services and platforms, connection testing is essential to verify connection parameters and configurations.
- Test for real-time systems: it is implemented on par with the previous test, verifying response times and possible delays, asynchronous transmissions and network outages (Pressman, 2010) .

CHAPTER III METHODOLOGY

For Caballero (2014), methodology is the "science whose specialty or field of study are the rational orientations that we require to solve new problems, and to acquire or discover new knowledge from those provisionally established and systematized by humanity" (p 78). on the other hand, for Hernández-Sampieri et al., (2018), methodology is also a reflection on the methods and techniques, which will help the researcher to achieve the objectives of the study,

In other words, the methodology is formed from a main plan constituted by the description of the units of analysis, the techniques of observation of phenomena, data collection and administration, the instruments and means of data collection, the procedures and techniques of data analysis (Baena Paz, 2017) , in other words it forms the general skeleton of the research that will later serve to articulate all the parts of the study, using the appropriate methodology, the objectives can be satisfactorily achieved.

This section describes the main research design processes, approach, scope, the paradigms used in its execution, its methodology and the tools used in data collection, as well as the validation techniques of the instruments and the elaboration of questionnaires.

3.1 Research approach

The quantitative route, as it is called by Hernández-Sampieri et al., (2018), is appropriate when you want to measure phenomena appropriately, or when you want to estimate dimensions and at the same time test hypotheses, in another work Hernández-Sampieri and Mendoza Torres (2018), argue that the quantitative approach is based on a deductive and logical scheme that seeks to formulate research questions and hypotheses to subsequently test them.

In this document, an applied type of research is addressed through a quantitative approach, as it is the one that is more in line with the research objectives, i.e. the use of Bot and its effects or incidences in relation to the treatment of smoking, from the results obtained through collection techniques such as the questionnaire or the survey, the proposed hypothesis is contrasted to determine its validity or not.

3.2 Scope of the research (research level)

The study of the present document is considered mainly of the exploratory type because it can lay the groundwork for subsequent descriptive, correlational or explanatory studies, in addition to this, according to Hernández-Sampieri et al, (2018), investigate phenomena or topics that have been studied little or not at all, of which there is no complete theoretical support, there are doubts about their scope or they have not been addressed in the context in which they appear, they also identify concepts with variables and hypotheses with possibly unexpected results, in fact they prepare the ground for broader, more elaborate and in-depth studies.

Another renowned author, Caballero (2014), argues that the most elementary level of the types of research is exploratory, the most common type of analysis is qualitative, but they can make references and give them a focus to data with quantitative precisions.

As mentioned by Hernández-Sampieri et al., (2018), a research can be identified as exploratory, descriptive, correlational or explanatory, but not strictly defined in one category, so sometimes it can belong in a sense to one, two or more categories of study

Therefore the present work is also considered correlational because its type of analysis is predominantly quantitative, using qualitative interpretations, estimating a relationship of a variable and its behavior with the effect and behavior of the others (Caballero, 2014) , in other words, this type of study has as its main objective to investigate the existing relationship between two or more variables speaking of a previous established context, and if this relationship is positive or negative it will be given the corresponding interpretation.

3.3 Research Design

The research design refers to the methods and techniques selected by the researchers (Hernández-Sampieri et al., 2018) , which by using them in a logical way will address the research problem in a more efficient amanera, this part of the research is made up of a series of steps on the How? Carrying out the research and how to prepare it, for Hernández-Sampieri et al., (2018), there are two categories in the design, experimental and non-experimental.

In the present study, two types of design are addressed, the first within classified as experimental, called pre-experimental, of the case study type of a single measurement, the second involves another of the non-experimental type named Transversal of the Exploratory type, categorized as shown in Table 1.

Table1 *Designs implemented in the present investigation*

Quantitative Designs	Group	Sub group
Experimental	Pre experimental	Case study with a single measurement
Non-experimental	Transversal	Exploratory

Note: Prepared by the authors.

For Arias González and Covinos Gallardo (2021), as negative points, the pre-experiment has no scientific value and does not guarantee causality, but on the other hand, it allows solving situational problems, and the authors also consider that a pre-experiment should contain the following characteristics: the groups are previously formed, there is only one experimental group, a pre-test and post-test can be applied, measurements can be taken at no more than two different times.

To perform the measurements, Arias González and Covinos Gallardo (2021) identify two scenarios, the first applies to studies of a group with a single measurement, the measurement is performed after applying the treatment in different periods of time. The second applies to studies of a group with two measurements, one before and one after the treatment, this is performed in three different time lapses.

In the present study, scenario one is used, denominated by Hernández-Sampieri et al., (2018) as "case study with a single measurement", which consists of providing a treatment to a group, to then apply the corresponding measurement of one or several variables, further arguing that this type of design does not meet the requirements of a pure experiment. Table 2 shows the design and the assignment of activities:

Table2 *Single experimental group design scheme with post-treatment measure, no pre-treatment measure.*

Groups	Subjects	Pre-treatment measure	Experimental treatment	Post-treatment measure
1	N	Not applicable	Y	X2

Note: Prepared by the authors

The second design associated with the present research has to do with non-experimental methods in their cross-sectional or trans-sectional type and in turn categorized as exploratory. According to Arias González and Covinos Gallardo (2021), cross-sectional designs collect data at a single point in time and only once.

For Hernández-Sampieri et al., (2018), in the cross-sectional design refers to taking a picture of something that happens, this type of studies can then have exploratory, descriptive and correlational scopes, thus supporting the correlational type scope mentioned in the previous section, also talking about the subgroup called exploratory, these have as main purpose to start with the study of variables selected at a specific time

To fulfill the general purpose of trans-sectional or cross-sectional designs, they must meet the following characteristics:

1. To describe variables in a group or, to determine what is the level of variables at a given time.
2. Evaluate a situation, community, event, phenomenon or context at a point in time.
3. Analyze the incidence of certain variables, their interrelation at a given time.

3.4 Variables and their operationalization.

For Arias Gonzales and Covinos Gallardo (2021), variables are that which will be studied, measured, controlled or manipulated, in addition they will be included in the title of the research, in the general objective, in the general problem and in the general hypothesis, on the other hand the author Baena Paz (2017), expresses that variables are instruments of analysis that make up the categories at a manifest level of reality classifying them into independent and dependent variables. Among other definitions, variables are elements that interact as causes and effects within the research process and are an integral part of the composition of experiments (C. E. E. E. E. E. Freire, 2018) .

The independent variable (x) is the characteristic or property that is supposed to be "the cause" of the studied phenomenon that cannot be controlled and the dependent variable (y) is the one whose modalities or values are related to the changes of the independent variable, i.e. "the effect" and that in turn can be scientifically controlled (Baena Paz, 2017) , for Hernández-Sampieri et al, (2018), the concept of variable applies to people, living beings, objects, facts and phenomena, which take different values with respect to the referred variable

The variables involved in this study are presented below:

y. Consumption reduction rate, associated and expressed through the effectiveness of treatment management processes. Dependent - effect

x. Bot for mobile devices. Independent - cause

As can be seen, "y" depends on the effects or changes that may occur with the implementation of "x". Table 4 shows the operationalization matrix of variables, their dimensions and indicators, as well as the number of items for each variable.

The operationalization of variables is used when it is required to define or conceptualize a variable, in the process it will go from being an abstract concept to one that can be quantified, thus defining its dimensions or values that it can take, in order to easily retrieve the information (C. E. E. E. E. Freire, 2018) . As a complement, Arias González and Covinos Gallardo (2021), argue that "the operationalization of variables is a process that occurs only in the quantitative approach because the variables must be susceptible to be observed and measured" (p. 60)

Operationalization will allow the translation of the theoretical variable into measurable and observable properties, from the general to the particular (Medina Martínez, 2015) . Medina (2015) defines operationalization as the process by which a composite theoretical variable is transformed into empirical variables that are easier to interpret and measure. For the author, operationalizing means identifying the variable, its dimensions and indicators, also including its conceptual and operational definition; these terms will be addressed in the following paragraphs

Returning to the conceptual definition, also called constitutive definition, for Hernández-Sampieri et al. (2018), it is the one that defines a variable in the context of the research, this definition must be validated by the scientific community and emerged from the literature review. Arias Gonzales and Covinos Gallardo (2021), agree that the conceptual definition is encompassed in the context of the research and that it is also taken into account from the population and the space where it is performed, they also argue that this definition should be different from the one established in the theoretical framework.

The operational definition is a set of activities that are carried out after the theoretical and practical analysis of the variables. This is done in order to establish how the variables will be measured, in other words, the operational definition allows to know which instrument or tool should be used to obtain clear and accurate results of the variable (Arias Gonzáles and Covinos Gallardo, 2021) .

Table 3 shows the conceptual definitions of the variables involved in this research, which were developed based on the opinions of the authors cited in this document and which refer to the subject in question.

Table3 *Conceptual Definitions of the variables*

Variables	Conceptual Definition
Consumption Reduction	Preventive and corrective procedures to eradicate or reduce tobacco use in persons classified as having a smoking problem.
Bot for mobile devices	A Bot is a software program programmed to perform fixed, repetitive tasks and interact with humans through text commands.

Note: Prepared by the authors

The dimensions are defined as the subdivided characteristics of the variable, (E. E. E. E. Freire, 2019) , the dimensions are raised taking into consideration the context of the research, as well as in the conceptual definition of the variable (Arias Gonzáles and Covinos Gallardo, 2021) .

Indicators are the properties of the variable subject to be measured (E. E. E. E. Freire, 2019) , in other words, they are elements determined within the dimensions and express the measurable reality of the variable (Baena Paz, 2017) .

Table4 *Matrix of operationalization of variables*

Variables	Dimensions	Indicators	Scale	Items
Consumptio n Reduction	Treatment	Time management	Ordinal	9
		General progress		

		Consumption reduction per day		
	Therapy	Successful therapies		
		Zero therapies		
	Logbook Management	Filling of logs		
		Logbook loss		
Bot for mobile devices	Operation	Program installation	Ordinal	13
		Search for contacts		
		Graphical interface appearance		
		Frequency of information received		
	Performance	Response time		
		Clarity of information		
		Ease of use		
		Data consumption		
		Speed of execution		
	Information	Help within the Bot		
		Tips received		

Note: Prepared by the authors

3.5 Hypothesis statement.

For Baena Paz (2017), operationally the hypothesis is a tentative answer to the research question, in this sense a hypothesis could be considered as a provisional proposition, i.e. an assumption to be verified, other authors such as Hernández-Sampieri et al., (2018), consider that "hypotheses indicate what we are trying to prove and are defined as tentative explanations of the investigated phenomenon" (p. 104).

In another contribution, the hypotheses are tentatively explained the research problem written as statements which subsequently support the line of study (Hernández-Sampieri and Mendoza Torres, 2018) . When formulating the hypotheses, the following guidelines can be followed to verify that they are written correctly:

- They should be stated as a specific statement.
- Your concept must be clear.
- That can be verified using instruments.
- It must be specific in tangible scenarios and with a logical limitation.
- They should be testable with research techniques. (Baena Paz, 2017) .

Following the above considerations, the hypotheses of the present research, known as research hypotheses or working hypotheses, are set out below, referring to the fact that the efficacy of the treatment will be directly related to the processes associated with the treatment against smoking:

HG. The use of a Telegram Bot increases the impact on the management of patients in treatment with smoking.

As an additional complement, null hypotheses can also be described, which are the opposite part of the research or working hypotheses, they are used to deny the statement of the research hypotheses, following with this logic, there can be as many null variables as research variables (Hernández-Sampieri et al., 2018) , in this sense the null hypotheses of the present work are established as follows:

Ho. The use of a Telegram Bot does not increase the impact on smoking treatment management.

To the previous series of hypotheses can also be added the so-called alternative hypotheses, which are alternative scenarios to those exposed by the research hypotheses and the null hypotheses, offering explanations different from those postulated. Alternative hypotheses can be formulated when other possibilities really exist, otherwise they should not be established (Hernández-Sampieri and Mendoza Torres, 2018) . The following are established as alternative hypotheses in the present work:

Ha. The use of a Telegram Bot partially increases the impact on smoking treatment management.

Ha1. The use of a Bot for mobile devices allows a semi-automated assessment of the viability of patients to verify their suitability for smoking cessation treatment.

Ha2. The acceptance rate of using a Bot for mobile devices for the management of smoking cessation treatment is significantly acceptable.

Ha3. The level of effectiveness in anti-smoking treatment is significantly reduced with the integration of a Bot for mobile devices.

Ha4. The rate of acceptance of the use of a web platform for the management of smoking cessation treatment is significantly acceptable.

3.6 Population

To make sense of the study of the present work, a group called population is required, with which through certain procedures the information is obtained which will be processed later, Hernández-Sampieri et al, (2018), defines it as; "set of all cases that agree with certain specifications" (p. 199), these specifications will be a guide for the selection of possible candidates, for Arias González and Covinos Gallardo (2021), population refers to "the population is the totality of elements of the study, it is delimited by the researcher according to the definition that is formulated in the study" (p. 113), likewise argues that the word population in terms of research methodology is synonymous with universe.

3.7 Sample

The sample represents the set of people, individuals or things that were selected to carry out a study, for the present work a non-probabilistic sample was carried out, also called directed. A non-probabilistic sample is a selection that is not subject to probability, but to the characteristics and context of the research, argued and supported by the researcher, the selected samples also meet other criteria (Hernández-Sampieri and Mendoza Torres, 2018) , the sample under study in the present work met the criteria established in the questionnaire "reasons for tobacco use", see Annex 1.

Of the non-probabilistic sample type, the quota sampling subtype was used, characterized by selecting subjects who share common characteristics within the population group or universe (Arias Gonzáles and Covinos Gallardo, 2021) . The inclusion criteria taken into account are mentioned below:

- Be over 18 years of age.
- To voluntarily accept to participate in treatment in accordance with the Juvenile Integration Center's own regulations.
- That they have applied the questionnaire "Reasons for tobacco consumption", obtaining a score determined as valid by the medical staff, with which a patient is considered suitable to undergo treatment.
- Have or have access to a smart mobile device with any of the following operating systems; Android or IOS, such device must have internet connection when you want to perform any operation related to the treatment.

The exclusion criteria, to determine which subjects do not meet the characteristics defined by the investigator, are as follows:

- They have applied the questionnaire "Reasons for tobacco use", obtaining a score determined as invalid by the medical staff, which is considered that a patient is not suitable to undergo treatment.
- Participants who do not have or do not have access to a smart mobile device with Android or IOS operating systems.
- Not be older than 18 years old.
- Failure to voluntarily accept to undergo treatment

3.8 Data Collection

Collecting data means applying one or more measurement instruments to collect relevant information on the study variables in the selected sample or cases (individuals, groups, organizations, processes, events, etc.). The data obtained are the basis of the analysis. Without data there is no research. However, to have reached this stage in the quantitative route, the hypotheses of the study and the variables, both conceptually and operationally, must have been established and defined with precision and clarity beforehand. Likewise, in the literature review, instruments or ways to measure or evaluate the variables proposed had to have been detected (Hernández-Sampieri and Mendoza Torres, 2018) .

3.8.1 Data Collection Technique

The survey is a tool that is carried out by means of an instrument called a questionnaire that will be detailed in later paragraphs, it is considered within the category of data collection techniques, it is focused on people and collects data related to their opinions, behaviors or perceptions of reality. The survey has quantitative or qualitative results and is composed of questions in a logical order established by the researchers (Arias Gonzáles and Covinos Gallardo, 2021) .

With the survey, mostly numerical data are obtained, it is a technique used quite regularly in social research and over time in scientific research. It is considered that every person has participated or will participate in a survey in some part of his or her life (López-Roldán and Fachelli, 2015) .

In this research work, the survey technique was used to collect data generated by patients and therapists through four questionnaires designed and adapted with the help of professionals in the addiction area of the Zacatecas Juvenile Integration Center; these questionnaires were analyzed using Cronbach's Alpha to verify their validity.

3.8.2 Data collection instruments

The questionnaire is perhaps the most widely used instrument for data collection, a questionnaire consists of a set of questions referring to one or several variables (Bourke et al., 2016) . It is the instrument par excellence in questioning techniques, it is important to consider the wording and position of the questions selected within the questionnaire (Baena Paz, 2017) .

In this thesis, a total of four questionnaire-type instruments were used to collect the data associated with this research. These questionnaires were designed, endorsed and validated to ensure their reliability, which were named as follows and are also available in the Annexes section of this document;

a) Tobacco use motives questionnaire
b) Bot Usability Questionnaire
c) Treatment efficacy
d) Usability of the web platform

The questionnaires used in this research are considered self-administered, as they were provided directly to the participants, in addition to being closed-ended questionnaires since they must respond to previously established answers and polytomous because they have three or more alternatives (Arias Gonzáles and Covinos Gallardo, 2021) .

3.8.3 Scaling

In the use of questionnaires, various types of questions are implemented, such as direct, closed, semi-closed or open questions. In the questionnaires implemented in the present study, closed questions are used in their entirety, so the best convenience is to use a scale that measures people's attitudes. The measurement of attitudes is used in various fields that are susceptible to research because they are useful for measuring perceptions of various qualities.

Attitudes are related to the personal perception of each individual and is influenced by previous knowledge or experiences, one of the best known methods to measure by scales the variables that constitute attitudes are: the Likert scaling method, the semantic differential and the Guttman scale (Hernández-Sampieri and Mendoza Torres, 2018) .

The Likert scaling method, used in this research, was developed by Rensis Likert in 1932, and is considered a popular method that is still in use today. It consists of a set of items or questions designed to capture the perception of the participants, to which the participants are asked to react. In other words, each statement is presented to the subject and he/she is asked to express his/her reaction by choosing one of the five points or categories of the scale. Each option is assigned a numerical value and at the end the total score is obtained by performing the corresponding sums (Hernández-Sampieri and Mendoza Torres, 2018) .

In other words, this method implements a rating scale that is used to question a person about his or her level of agreement or disagreement of a given situation, thereby knowing the degree of conformity or non-conformity

Subjects respond specifically based on their level of agreement or disagreement, Likert frequency scales use fixed and closed response format that are used to measure attitudes and opinions. These scales allow determining the level of agreement or disagreement of respondents

Responses may contain different levels of measurement such as 5, 7 and 9 items, determined as an odd scale, or 4, 6 and 8 items determined as an even scale.

The scales with odd numbers offer a neutral opinion option to the participants, the scales with even numbers allow to choose between positive or negative answers without a middle point of selection, for the questionnaires used in the present investigation it was decided to use a set of even answers, to force the respondent to select an option of agreement or disagreement instead of taking a neutral position, Table 5 shows as an example the design of one of the questions of the questionnaire "Reasons for tobacco use":

Table5 *Example of question design using Likert scaling method*

Ask	Possible answers	Value assigned
I feel that smoking gives me peace of mind	Never	1
	Rarely	2
	Occasionally	3
	Followed by	4
	Frequently	5
	Always	6

Note: Prepared by the authors

3.8.4 Determining instrument reliability

There are numerous procedures to calculate the reliability of a measuring instrument, most of them use formulas that determine a reliability coefficient expressed in values ranging from zero to one, where zero means zero reliability and one represents the maximum reliability, the closer the value is to zero, the greater the error in the measurement will be.

Hernández-Sampieri et al., (2018), consider that the most commonly used procedures to determine the reliability of an instrument are the following: a) measure of stability (test-retest reliability), b) method of alternative or parallel forms, c) split-halves method (split-halves) and d) internal consistency measures. Some authors consider that reliability coefficients above 95% may imply redundancy of items or items (Hernández-Sampieri and Mendoza Torres, 2018) .

In this case, the method of internal consistency measures was selected to validate the instruments of the present study, through the Cronbach alpha coefficient.

The questionnaires used in this study were validated by means of Cronbach's alpha under the SPSS statistical analysis program, which works by means of internal consistency measures, which allows estimating the reliability of a measurement instrument through a set of items that are expected to measure the same construct or theoretical dimension. The coefficient measured by Cronbach's alpha, assumes that the items implementing a Likert scale, measure the same construct and that they are highly correlated (Frías-Navarro, 2022) .

As general criteria, Gliem and Gliem (2003) suggest the following recommendations for evaluating Cronbach's alpha coefficients:

- Coefficient alpha >.9 is excellent.
- Coefficient alpha >.8 is good.
- Coefficient alpha >.7 is acceptable.
- Coefficient alpha >.6 is questionable.
- Coefficient alpha >.5 is poor

- Coefficient alpha <.5 is unacceptable.

3.8.4.1 Reliability results of the instruments used.

The reliability of a measurement instrument used in research refers to the degree to which its repetitive application leads to the same results (Hernández et al., 2016) , reliability refers to the consistency or stability of a measurement. A technical definition of reliability that helps to solve both theoretical and practical problems is that which starts from the investigation of how much measurement error exists in a measurement instrument, considering both the systematic variance and the variance due to chance (Fred and Howard, 2002) .

Based on the above, the results obtained regarding the reliability of the instruments, calculated by means of the SPSS statistical program, are shown below.

3.8.4.2 Analysis of the questionnaire "Motives for tobacco use".

For questionnaire 1 called "Motives for tobacco use" (see Annex 1), the following results were obtained when performing the Cronbach's alpha analysis, and also the 60 records were taken into account without excluding any, it should be remembered that the present questionnaire was used to determine whether a patient is suitable to take the treatment against smoking, Table 6 refers to the number of cases involved in this case 60, it also refers to the fact that no item was excluded from the analysis.

Table6 *Summary of case processing of the questionnaire "Motives for Tobacco Use".*

		N	%
Cases	Valid	60	100.0
	Excluded	0	.0
	Total	60	100.0

a. Elimination by list is based on all the variables in the procedure.

Note: Prepared by the authors.

Table7 *Cronbach's alpha of the questionnaire "Motives for tobacco use".*

Reliability statistics	
Cronbach's alpha	N of elements
.885	60

Note: Prepared by the authors.

Taking into account the values obtained in Table 7, the minimum acceptable value for Cronbach's alpha coefficient is 0.70; a lower value represents low internal consistency. On the other hand, an ideal value would be 0.90, a higher value is considered with redundancies or duplications (Oviedo and Campo-Arias, 2005) .

As can be seen in Table 8, all questions have an individual Cronbach's Alpha above 0.8 and close to 0.9, which indicates that all questions are reliable and, furthermore, the elimination of one question from the questionnaire would not affect the final reliability of the questionnaire.

Table8 *Statistics on suppression of items of the questionnaire "Motives for tobacco use".*

Total element statistics				
	Average of scale if the element has been suppressed	Scale variance if the element has been suppressed	Total correlation of corrected elements	Cronbach's alpha if the element has been deleted

Question1	85.1667	350.209	.538	.879
Question2	84.2500	343.614	.539	.879
Question3	84.6500	340.875	.636	.877
Question4	84.9833	349.271	.566	.879
Question5	84.7000	346.383	.493	.880
Question6	84.0167	341.678	.604	.877
Question7	84.4833	330.830	.731	.874
Question8	84.9833	364.898	.228	.884
Question9	84.4833	364.051	.247	.884
Question10	84.4500	355.947	.520	.880
Question11	84.6000	352.346	.370	.882
Question12	85.1500	370.469	.044	.888
Question13	85.2000	364.366	.176	.886
Question14	84.9667	351.660	.503	.880
Question15	84.7000	349.366	.452	.881
Question16	84.1667	340.921	.552	.878
Question17	84.6500	361.282	.255	.884
Question18	84.6667	354.802	.421	.881
Question19	84.6000	360.312	.226	.885
Question20	84.8333	356.514	.366	.882
Question21	84.9333	362.673	.235	.885
Question22	84.3167	356.525	.336	.883
Question23	84.5667	361.707	.239	.885
Question24	84.8833	359.020	.331	.883
Question25	84.5167	356.051	.325	.883
Question26	84.1500	356.333	.324	.883
Question27	84.9833	348.084	.546	.879
Question28	84.9000	362.736	.244	.884
Question29	84.4500	361.031	.247	.884
Question30	84.3667	360.440	.254	.884
Question31	84.8500	362.435	.175	.887
Question32	84.7167	348.105	.489	.880
Question33	84.5000	346.695	.506	.880
Question34	84.7000	340.993	.648	.877
Question35	85.1333	350.897	.513	.880

Note: Prepared by the authors

It is worth mentioning that the answers to the previous questionnaire were modified with respect to the original published in the document "Tratamiento para dejar de Fumar" elaborated by Centros de Integración Juvenil page 28 , in the present document it is marked as Annex 2.

The original questionnaire only handles 3 options for answers, while the one used in the present research covers six options, giving a wider range of decision for each patient and eliminating the neutral option, to force users to opt for negative or positive opinions in a closed way, this change was recommended by the people who managed the treatments and the questions involved had no changes.

The answers to the original questionnaire are as follows:

1. If you "very frequently" smoke for that reason
2. If you "occasionally" smoke for that reason
3. If you "never" smoke for that reason

Those used in this document are:

1. If you "Never" smoke for that reason
2. If you "Rarely" smoke for that reason
3. If you "Occasionally" smoke for that reason
4. If you "Seguido" smoke for this reason
5. If you "Frequently" smoke for that reason
6. If you "Always" smoke for this reason

3.8.4.3 Analysis of the questionnaire "Bot Usability".

The reliability results obtained from the Cronbach's alpha of questionnaire 2 called "Bot Usability Questionnaire" (See Annex 2), are shown in Table 9, which shows the reliability of the instrument used to measure the usability of the Bot, in addition, all 60 records were taken into account without excluding any, it should be remembered that this questionnaire was used to determine the level of usability of the Bot that was used as a support in the treatment against smoking.

Table9 *Summary of case processing of the questionnaire "Bot Usability".*

		N	%
Cases	Valid	60	100.0
	Excluded	0	.0
	Total	60	100.0
a. Elimination by list is based on all the variables in the procedure.			

Note: Prepared by the authors.

As shown in Table 10, all the questions have a Cronbach's Alpha above 0.9 and close to 0.95, which indicates that all the questions are reliable; the addition or elimination of a question from the questionnaire would not affect the final reliability of the questionnaire

Table10 *Cronbach's Alpha "Bot Usability Questionnaire".*

Reliability statistics	
Cronbach's alpha	N of elements
.952	13

Note: Prepared by the authors

The Cronbach's Alpha coefficient applied to the items of the instrument was calculated with SPSS software and its result is 0.952, which according to the interpretation of Oviedo and Campo (2005), there is "redundancy or duplication", because it is in the range 0.90-1. Therefore, it is concluded that the internal consistency of the instrument used could have some duplications in the proposed items, without affecting its viability. Table 11 shows the statistics corresponding to each of the items of the "Bot Usability" questionnaire.

Table11 *Statistics on suppression of elements*

Total element statistics				
	Average of scale if the element has been suppressed	Scale variance if the element has been suppressed	Total correlation of corrected elements	Cronbach's alpha if the item has been deleted
Question1	56.8500	217.452	.677	.950
Question2	55.5833	207.603	.923	.943
Question3	56.2333	211.979	.762	.948
Question4	55.9500	212.353	.822	.946
Question5	56.2667	213.284	.753	.948
Question6	55.8833	209.630	.909	.944
Question7	57.1000	218.668	.437	.961
Question8	55.7500	209.852	.905	.944
Question9	56.1167	212.037	.850	.946
Question10	56.2833	212.817	.857	.945
Question11	56.1833	212.356	.832	.946
Question12	57.1333	220.287	.559	.954
Question13	56.2667	214.504	.768	.948

Note: Prepared by the authors.

3.8.4.4 Analysis of the "Treatment Effectiveness" Questionnaire

Results obtained from questionnaire 3 "Treatment efficacy" (See Annex 3), Tables 12 and 13 show the Cronbach's alpha value for the questionnaire used to measure treatment efficacy and its items, respectively. This questionnaire has the lowest Cronbach's alpha value because it was completed exclusively by therapists who in their eagerness to maintain the privacy of patient data may have biased the responses, even so, the Cronbach's alpha value of the questionnaire is reliable since it is in the range considered as good.

Table12 *Cronbach's Alpha "Treatment Effectiveness Questionnaire".*

Reliability statistics	
Cronbach's alpha	N of elements
.825	9

Note: Prepared by the authors

The obtained results described in Table 13 yielded a Cronbach's Alpha above 0.8, of the total number of items listed, only three are below 0.8 if the item in question were to be removed, indicating that the reliability is in the range of Acceptable and Good to between the range of 0.8 to 0.9

Table13 *Statistics on suppression of elements*

Total element statistics				
	Scaling average if the element has been suppressed	Scale variance if the element has been suppressed	Total correlation of corrected elements	Cronbach's alpha if the item is deleted
Question1	23.2667	69.623	.605	.798
Question2	22.9167	70.722	.548	.805

Question3	22.9667	74.168	.361	.827
Question4	23.2333	67.436	.589	.800
Question5	22.7333	66.945	.591	.799
Question6	23.0333	69.660	.514	.809
Question7	23.2833	71.766	.556	.805
Question8	23.2167	67.868	.596	.799
Question9	23.4833	78.695	.414	.820

Note: Prepared by the authors.

Table 14 refers to the number of cases involved in this case 60, it also refers to the fact that no element was excluded from the analysis.

Table14 *Summary of case processing of the "Treatment Effectiveness" questionnaire.*

		N	%
Cases	Valid	60	100.0
	Excluded	0	.0
	Total	60	100.0
a. Elimination by list is based on all the variables in the procedure.			

Note: Prepared by the authors

3.8.4.5 Analysis of the questionnaire "Usability of the Web platform".

The results obtained from questionnaire 4 "Usability of the Web platform" (See Annex 4), are shown in tables 15 and 16, where the Cronbach's alpha value of the complete questionnaire of the tendency to eliminate one item is expressed, this questionnaire was solved exclusively by therapists who in their eagerness to maintain the privacy of the patients' data could have biased the answers, Even so, the Cronbach's alpha value of the questionnaire is reliable since it is in the range considered as good. Table 15 refers to the number of cases involved in this case 5, it also refers to the fact that no item was excluded from the analysis.

Table15 *Summary of case processing of the questionnaire "Usability of the Web platform".*

		N	%
Cases	Valid	5	100.0
	Excluded	0	.0
	Total	5	100.0
a. Elimination by list is based on all the variables in the procedure.			

Note: Prepared by the authors.

Table 16 shows the Cronbach's Alpha value above 0.85 and close to 0.90 which indicates that all the questions are reliable, the elimination of one question from the questionnaire would not affect the final reliability of the questionnaire.

Table16 *Cronbach's Alpha "Web platform usability questionnaire".*

Reliability statistics	
Cronbach's alpha	N of elements
.908	10

Note: Prepared by the authors

Table 17 shows the statistics corresponding to each of the items of the "Bot Usability" questionnaire, and the internal consistency measure for each of the questions.

Table17 *Statistics on suppression of elements*

Total element statistics				
	Scaling average if the element	Scale variance if the element	Total correlation	Cronbach's alpha if the item has

	has been suppressed	has been suppressed	of corrected elements	been deleted
Question1	40.2000	68.700	.853	.886
Question2	40.6000	75.800	.477	.913
Question3	40.0000	80.000	.386	.915
Question4	39.8000	69.200	.761	.893
Question5	39.6000	81.300	.394	.913
Question6	40.0000	81.500	.404	.912
Question7	39.6000	72.300	.882	.887
Question8	39.8000	71.700	.917	.885
Question9	39.0000	72.500	.743	.894
Question10	39.2000	69.200	.982	.880

Note: Prepared by the authors

3.8.5 Determination of instrument validity

Validity determines the level at which an instrument measures one or several variables that are intended to be measured, it also analyzes whether the instrument really defines the concept of the variable by means of its indicators. Validity is a characteristic that every measurement instrument should achieve (Hernández-Sampieri and Mendoza Torres, 2018) . At this point validity would have to answer the question; it measuring what it thinks it is measuring, if the answer is affirmative, the instrument measuring said variable is valid, (Kerlinger, 1979) . It is also necessary to check if and how other researchers have measured this variable, in order to select the items carefully.

To validate the instruments used in the present research, the method called construct validity was used, which consists of taking the data obtained to establish the relationships between the variables under study and thus determine the existence of possible constructs that support the design of the instruments these procedures are performed with multivariate statistical techniques (Pulido, 2018) .

The aforementioned procedures allow the variables and their data to be organized, delimited and presented in a more accessible visual structure. The method used to determine the number of factors and the nature of a group of underlying constructs in relation to a set of measurements referring to a group of variables is known as exploratory factor analysis.

In order to perform a factor analysis, there are several tests available, among which the following stand out: determinant of the correlation matrix, Bartlett's sphericity, the Kaiser Meyer and Olkin index (KMO). That said, some authors suggest performing at least two of the above tests, to determine whether any of them shows some degree of correlation and consider that it makes sense to perform the analysis (Martínez and Sepúlveda, 2012) .

In this study it was determined to use Bartlett's Sphericity test and the Kaiser-Meyer-Olkin (KMO) test to determine the validity of the instruments used. Bartlett's Sphericity test determines by means of high values of x2 (chi-square) its significance value, this implies the existence of high correlations, this test is significant if the values obtained are less than 0.000 or otherwise 0.050 or 0.010, with a confidence level of 95% and 99%, respectively (Crismán-Pérez and Núñez-Vázquez, 2015) .

The KMO test is a measure of the suitability of data for factor analysis. The test measures the sampling fit for each variable, the statistic used is a measure of the proportion of variance between variables that could share a variance. The value obtained by means of the KMO test can vary between 0 and 1, obtaining a value close to zero does not support or support the factorial analysis, values below 0.70 are considered unfavorable, while values between 0.70 to 0.79 are considered regular, between 0.80 to 0.89 are considered meritorious, and from 0.90 to 1 wonderful (Caballo Trebol, 2013) .

Other authors admit and consider values between 0.6 and 0.69 as the minimum acceptable KMO values, values below 0.6 indicate that the sampling is not adequate and that corrective measures should be taken (Vogt and Johnson, 2015) . Others consider that the minimum value for KMO could be 0.5, and determine that each researcher should apply his or her own judgment if he or she encounters this range of values (Suárez, 2007) .

3.8.5.1 Results of the validity of the instruments used.

Following the recommendations of the previous section, the validity of the instruments was carried out by means of the SPSS statistical program through an analysis of dimensions, which are described below; for the "Motives for tobacco use" questionnaire, the following dimensions were determined: sensation, appreciation, instinct, behavior. For the questionnaire "Usability of the Bot" the following were considered; operation, performance and information. For the "Treatment Effectiveness" questionnaire the following were identified; treatment, therapy, log management and finally, for the "Web

Platform Usability" questionnaire the following dimensions were selected; interface, functionality, performance.

For the questionnaire "Usability of the Web platform ", the Kaiser-Meyer-Olkin (KMO) measures yielded a value of 0.574 considered as a minimum acceptable value and Bartlett's test of sphericity at its significance level was .000 considered as valid, both values showed that the samples met the criteria for factor analysis. The results of this study indicate that the instrument has good construct validity; these results are shown in Table 18.

Table18 *KMO values and Bartlett's test of sphericity questionnaire "Usability of the Web platform".*

KMO and Bartlett's test		
Kaiser-Meyer-Olkin measure of sampling adequacy		.574
Bartlett's test for sphericity	Approx. chi-square	131.585
	gl	45
	Sig.	.000

Note: Prepared by the authors.

For the "Bot Usability" questionnaire, the Kaiser-Meyer-Olkin (KMO) measures yielded a value of 0.603, exceeding the minimum acceptable value, and Bartlett's test of sphericity at its significance level was .001, considered valid , both values showed that the samples met the criteria for factor analysis. The results of this study indicate that the instrument has good construct validity; these results are shown in Table 19.

Table19 KMO values and Bartlett's test of sphericity questionnaire "Bot Usability".

KMO and Bartlett's test	
Kaiser-Meyer-Olkin measure of sampling adequacy	.603

Bartlett's test for sphericity	Approx. chi-square	121.465
	gl	78
	Sig.	.001

Note: Prepared by the authors.

For the questionnaire "Consumption motives", the Kaiser-Meyer-Olkin (KMO) measures yielded a value of 0.571, exceeding the minimum acceptable value, and Bartlett's test of sphericity at its level of significance was .000, considered valid; both values showed that the samples met the criteria for factor analysis. The results of this study indicate that the instrument has good construct validity, these results are shown in Table 20.

Table20 KMO values and Bartlett's test of sphericity questionnaire "Consumption motives".

KMO and Bartlett's test		
Kaiser-Meyer-Olkin measure of sampling adequacy		.571
Bartlett's test for sphericity	Approx. chi-square	1170.103
	gl	595
	Sig.	.000

Note: Prepared by the authors.

For the questionnaire "Consumption motives ", the Kaiser-Meyer-Olkin (KMO) measures showed a value of 0.846, exceeding the minimum acceptable value, and Bartlett's test of sphericity at its significance level was .000, considered valid; both values showed that the samples met the criteria for factor analysis. The results of this study indicate that the instrument has good construct validity; these results are shown in Table 21.

Table21 KMO values and Bartlett's test of sphericity questionnaire "Consumption motives".

KMO and Bartlett's test		
Kaiser-Meyer-Olkin measure of sampling adequacy		.846
Bartlett's test for sphericity	Approx. chi-square	145.406
	gl	36
	Sig.	.000

Note: Prepared by the authors.

Once this was done, we proceeded to evaluate the objectivity of each and every one of the instruments in each of their areas.

3.8.6 Determination of instrument objectivity

The term objectivity in relation to quantitative research, is related to a standard associated with the way in which phenomena are captured in reality, from the point of view of people it can be a complicated process to reach such understanding, sometimes it is achieved through consensus among researchers or by performing multiple measurements (Hernández-Sampieri and Mendoza Torres, 2018) .

3.8.6.1 Results of the objectivity of the instruments used.

Supported by the objectivity, experience and professionalism of the therapists, medical personnel and other health specialists involved in the treatment, it was determined that the instruments used in the present investigation fulfill their purpose of measuring, defining and evaluating the object to be evaluated, in addition to complying with the elements of control of said measurement, to identify the external variables that affect the results and that both the environment and those involved do not interfere with the results of the investigation.

3.9 Data analysis

Data analysis is the science of examining a set of data for the purpose of drawing conclusions about the information in order to make decisions, or simply to expand knowledge on various topics.

Data analysis consists of several tests and operations on the information collected, with the aim of obtaining precise conclusions that help to achieve the proposed objectives. An implementation with Likert-type scales support works to perform analyses for ordinal variables (non-parametric), interval (parametric) and at the end verify their coincidence (Hernández-Sampieri and Mendoza Torres, 2018) .

In the present work and supported by inferential statistics, the following parametric tests were performed; Pearson correlation coefficient and linear regression, to determine whether there is any positive, negative or neutral correlation between the variables exposed in this study, in addition to the basic tests of descriptive statistics, Hernandez-Sampieri et al, (2018) suggests, when Likert-type scales are used, analyses should be implemented for both ordinal and interval variables and then verify their coincidence, i.e. use Pearson's coefficient and Spearman's coefficient and contrast the results

Pearson's correlation coefficient is a statistical test used to analyze the relationship between two variables, it is calculated from the scores obtained through the instrument in a sample of two variables, depending on the result, Pearson's r coefficient can range from -1.00 to +1.00, Table 22 describes the different levels of correlation possible.

Table22 *Pearson's Correlation Levels*

Correlation level	Interpretation of correlation
-1.00	Perfect negative correlation ("The higher X, the lower Y" or vice versa).
-0.90	Very strong negative correlation
-0.75	Significant negative correlation
-0.50	Average negative correlation.
-0.25	Weak negative correlation.
-0.10	Very weak negative correlation.
0.00	There is no correlation between the variables.
+0.10	Very weak positive correlation.
+0.25	Weak positive correlation.
+0.50	Average positive correlation.
+0.75	Significant positive correlation.
+0.90	Very strong positive correlation
+1.00	Perfect positive correlation ("The greater X, the greater Y" or vice versa).

Note. Adapted from Hernández-Sampieri et al., (2018).

Linear regression is another statistical model used to estimate the effect of one variable on another; the higher the correlation, the better the prediction. For its interpretation, linear regression is described starting with the elaboration of a scatter diagram, which are useful to graphically visualize a correlation (Hernández-Sampieri et al., 2018) .

Finally, the correlation coefficients for ordinal variables of Spearman and Kendall ordered ranks are correlation measures for variables in an ordinal level of measurement, these coefficients are mostly used to statistically associate Likert-type scales, to analyze the results, the rs and t coefficients are used, both coefficients vary from -1.0 identified as a perfect negative correlation to+ 1.0 identified as a perfect positive correlation, taking 0 as absence of correlation, for intermediate values its significance is interpreted the same as Pearson's coefficient (Hernández-Sampieri and Mendoza Torres, 2018) .

3.9.1 Software for statistical analysis

In the present research work we used the statistical management software called SPSS developed by the IBM company, SPSS (Superior Performing Software Systems) is a complete statistical software package relatively easy to use, popular in areas such as economics and social sciences (Pedroza and Dicovskyi, 2007)

This program is compatible with several operating systems, is a software used to capture and analyze data and also in the creation of tables and graphs with complex data, it is used to perform descriptive statistical analysis, bivariate, regression, factor analysis, and graphical representation of data, among others. It was initially designed to work with research related to the social sciences, but has subsequently been implemented in a wide variety of research areas

CHAPTER IV ANALYSIS AND INTERPRETATION OF THE RESULTS

The results of any research are a fundamental part of it since they are the ones that give firmness and weight to the hypotheses with the sole purpose of validating them, the objective of the results section is to present all the important information obtained throughout the study, presenting a logical sequence in the writing and organization of the information, the researcher must report clearly and impartially the data obtained.

One of the primordial needs in all research is the requirement of a clear processing of the information in order to be able to interpret the investigated subject matter and achieve suitable results (Baena Paz, 2017) .

In this chapter and for the present study, the analysis of the results found with the collection of data through the Bot and some other technological tools by means of instruments specific to each segment, and then processed with SPSS software, is shown. This will be used to evaluate the usability of both the Bot and the support web platform, the effectiveness in the treatment against smoking and the relevance of whether a patient is suitable or not to take the treatment, in addition to this, correlation between the exposed variables will also be analyzed

In order to understand the context of the results to be presented, a brief recapitulation of the environment in which and how the study was conducted will be made. The treatment against smoking is a service provided by the

Juvenile Integration Centers, within the procedure there is a point in which patients record their consumption habits on sheets of paper as daily logs, but these are easily lost or forgotten and are an impractical method when carrying them with them everywhere, which is why the doctor in charge cannot keep track of the consumption habits of patients and therefore cannot determine a timely diagnosis and treatment.

A mobile device is a tool that most people can acquire and manipulate, even more so it is considered that they are familiar with instant messaging applications for being a massively used means nowadays for interpersonal communication, therefore, an application that is able to record daily smoking events within a messaging application, is highly useful and is an ideal complement to carry more effectively the treatment against smoking.

Before starting with the presentation of results, a brief tour will be made describing the most important parts of the Bot developed by Velázquez Macias et al., (2017), implemented as a support in the treatment against smoking.

4.1 Requirements functional requirements of the Bot

As main requirements for a correct operation of the Bot, the following issues are contemplated on the client or user side;

- Mobile device with a modern operating system based on Android, IOS, Windows Phone, or any modern web browser compatible with Windows, Linux or Mac.
- Telegram application installed.

- WIFI internet connection or via mobile data.

The following issues are considered as the main requirements on the server side;

- Windows, Linux or Mac operating system
- Correctly configured Python programming software.
- Code script written in Python running indefinitely on a local machine which listens for requests from clients and stores the records sent by them in a database to be consulted later.
- Permanent and stable internet connection.

4.1.1 Functionality Bot logic

Based on a client-server architecture the main application runs on a system capable of running applications written in the Python programming language implemented in the Telepot libraries, the source code of the Bot can be found in Appendix A of this document, the client-server interaction is achieved through text messages sent over the Internet through the Telegram application, Figure 5 shows the logical operation that follows the operation of the Bot.

Figure5 *Diagram on the logical operation supporting the Bot*

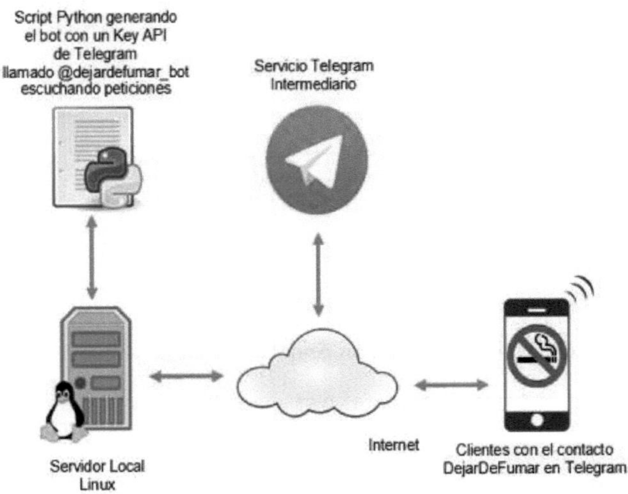

Note. Taken from (Velázquez Macias et al., 2017, p.x. 58).

The client-server interaction, the messages, their responses, and the body of the texts presented by Bot are based on the material of Centros de Integración Juvenil para el tratamiento de las adicciones (Juvenile Integration Centers for the treatment of addictions).

4.1.2 Execution and implementation of Bot

After adding as a contact the Bot called @dejardefumar_bot identified by the international signage of "no smoking", the first message is automatically sent from the user's phone to the Bot, that is from a client to the server, the default command when starting a Bot in Telegram is "/start", the server responds immediately with a first message, in this case on the server side with a welcome text, It is common to assign a proper name to the Bot to simulate a more natural chat, complementing the welcome message also includes a menu of options to help on the commands supported by the Bot, which is represented in Figure 6, these two elements serve as an instruction to the communication that will take place between the user or patient in treatment and the Bot.

Figure6 *Screenshot with welcome and help message*

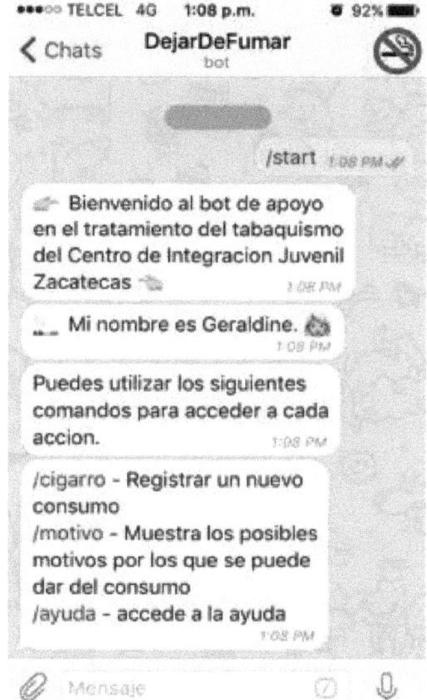

Note. Taken from (Velázquez Macias et al., 2017, p.x. 59).

Another of the main functions of the Bot, as shown in Figure 7, is to record the patient's tobacco consumption, giving the opportunity to select the reason for which a cigarette was lit, as well as some other data that help the medical staff to make certain decisions concerning the patient's treatment.

Figure7 *Screenshot of consumption records*

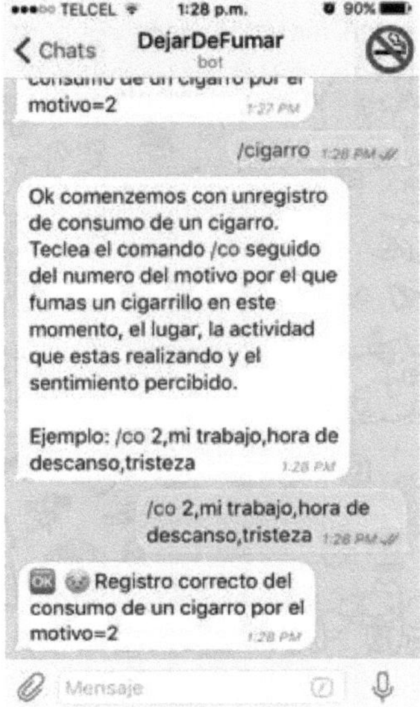

Note. Taken from (Velázquez Macias et al., 2017, p.x. 59).

When a record is received through the "/co" command, it is stored in a database hosted on the server, subsequently the records and their properties can be consulted by the therapist or specialized medical personnel. Table 23 below shows the format of records and their attributes and how they are stored in the database:

Table23 *Consumption record in the database*

Consumption_id	Id_chat	Date-time	Motive	Location	Activity	Feeling
1	22345	12/12/2017 12:23	2	Auto	Break time	Anger
2	24598	12/12/2017 14:23	3	House	After lunch	Sadness

Note. Adapted from Velázquez Macías et al., (2017).

Another of the main options represented in Figure 8, is to start the questionnaire that serves as a basis for the initiation of treatment against smoking, can be invoked with the option "/questionnaire", this questionnaire served as a basis for determining whether a patient is suitable for treatment according to the results obtained.

Figure8 *Screenshot of the start of the smoking motives questionnaire.*

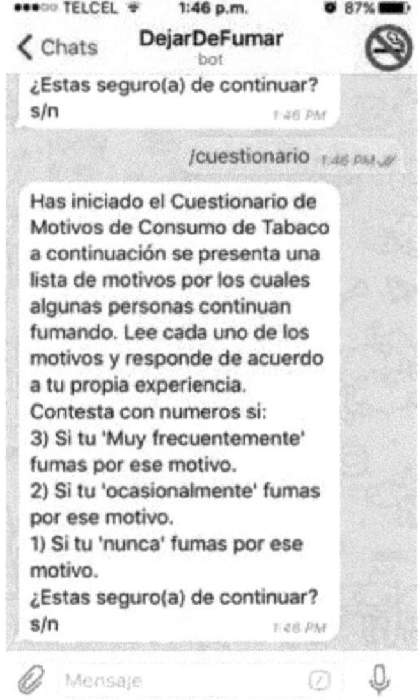

Note. Taken from (Velázquez Macias et al., 2017, p.x. 60).

As a reinforcement option towards the treatment, exemplified in Figure 9, it is shown how the Bot is able to send complementary tips authorized by health specialists who also argue that they can serve to avoid or reduce tobacco consumption, these messages are sent randomly and can be programmed with a certain frequency.

Figure9 *Screenshot with tips sent periodically from the server to each client.*

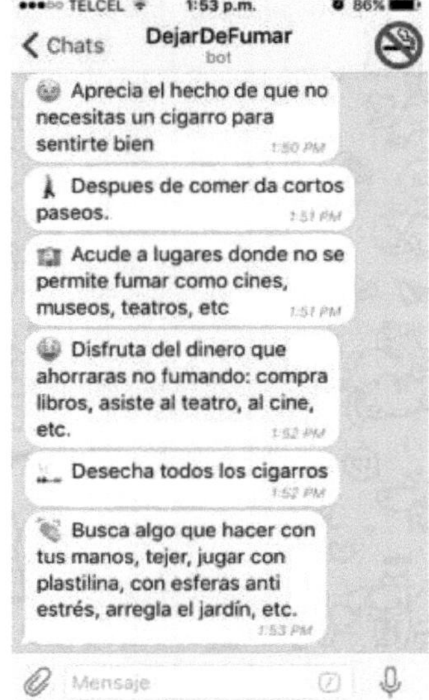

Note. Taken from (Velázquez Macias et al., 2017, p.x. 60).

These messages are validated according to Annex 3 of the Implementation Manual called "Treatment to Quit Smoking" developed Juvenile Integration Centers and the Directorate of Treatment and Rehabilitation

According to the experts' opinion, a greater number of tips can be configured from those available in the Bot, in order to increase the variety of information. Additionally and as part of the natural interaction process, the Bot responds to certain messages, for example; "good morning", "good evening", "thank you", among others.

When making mistakes in typing commands or sending unrecognized commands, the Bot has an error handling system which responds when the server does not know a command with an alternative text, as shown in Figure 10.

Figure10 *Screenshot with greeting and alternative messages*

Note. Taken from (Velázquez Macias et al., 2017, p.x. 61).

4.2 Operation of the consultation website

To support the consultation of patient records, a simple web site was designed and made available to therapists by assigning them user names and passwords so that they could provide personalized follow-up for each patient.

The individualized follow-up and interpretation of the records generated by the patients were the responsibility of the assigned therapist, so those results are beyond the scope of this work.

It should be noted that the names of all patients were associated with numerical identifiers in order to protect their identity and privacy.

By adding the contact @dejardefumar_bot in the Telegram messaging application, described in the previous section, the program algorithms assign a unique and unrepeatable chat identifier for each patient, this identifier named in the database as "id_chat" is assigned to the patient's file by the therapist, being the only one who knows to which person this identifier belongs.

Within the site, the therapist was able to perform various types of queries, among which the following stand out: search by chat identifier, search by place of consumption, search by sentiment, search by date and time of registration, among others, in addition to having the possibility of generating files according to the queries made in PDF format or Excel file format, besides being able to directly print the results for greater understanding and subsequent interpretation, the capture of the main screen of the Web interface that includes all the options it supports and shows all its graphic components which serve to support the treatment against smoking is shown in Figure 11.

Figure11 *Consumption Record query page screen*

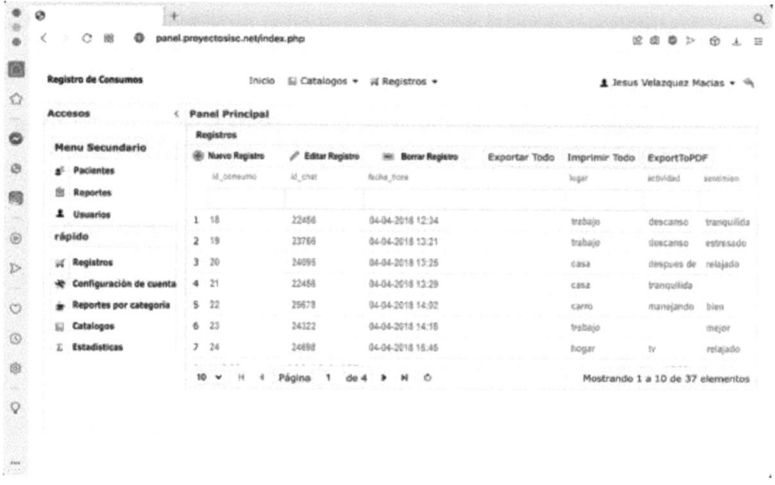

Note: Own elaboration.

Within the site, the therapist was able to perform various types of queries, including: by chat identifier, by place of consumption, by sentiment, by date and time of registration, among others, in addition to having the possibility of generating files according to the queries made in PDF format, or directly printing the results for greater understanding and subsequent interpretation.

This tool allowed the therapist to facilitate the diagnosis and the determination of the best treatment according to the characteristics and consumption behaviors of each patient, this information is of a private nature and is not available, so it is beyond the scope of this document.

4.3 Traditional method of work

The treatment against smoking is a service provided by the Juvenile Integration Centers, within the procedure there is a point in which patients record their smoking habits on sheets of paper as daily logs, but these are easily lost or forgotten and are an impractical method when carrying them with them everywhere, which is why the doctor in charge cannot keep track of the patients' smoking habits and therefore cannot determine a timely diagnosis and treatment.

A mobile device is a tool that most people can acquire and manipulate, more to a are familiar with instant messaging applications, therefore, an application that is able to record such consumption events within a messaging application is highly useful and is an ideal complement to more effectively bring your smoking treatment.

Data collection consisted of an initial questionnaire (See Annex 1) to determine whether or not a person is eligible to participate in the study, followed by another questionnaire to determine the effectiveness of the Bot.

4.4 Interpretation of the results obtained

This section shows the results obtained from the research methods and techniques described above, according to Baena Paz (2017), "the new needs of research require clear, understandable and effective information processing to be able to interpret the reality under investigation and have suitable results" (p. 110). For their part, Hernández-Sampieri et al. (2018), believe that "Once the data have been coded, transferred to a matrix, stored in a file and "cleaned" of errors, the researcher proceeds to analyze them" (p. 272).

4.4.1 Preliminary work for sample collection

In order to obtain the candidates to participate in the research, a questionnaire previously designed by Centros de Integración Juvenil called "Cuestionario de Motivos de consumo de tabaco" was used to determine if a person is first suitable to start a treatment, it should be noted that this selection was in charge of the therapeutic staff of the Centro de Integración Juvenil, as professionals trained in that area.

The Bot was used for patients to answer the questionnaire digitally and from their mobile devices, the goal was to gather at least 60 patients, this number was determined according to the records of previous years in which approximately is the number of people who take this type of treatments per year in that center, according to the opinion of the staff commissioned in that center.

Figure 12 shows the results obtained by each patient, which, in the opinion of the therapists and based on the operating manuals, determine whether or not they are candidates for treatment; an average range was established from 1.5 to 6.0 for viable candidates and from 1 to 1.49 for non-viable candidates:

Figure12 *Scatter plot of the average obtained per patient in the consumption motives questionnaire.*

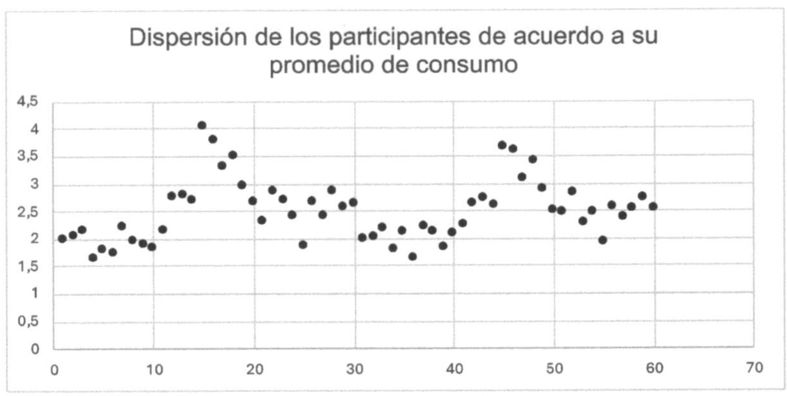

Note: Own elaboration (Velázquez, 2022).

Given these results, the trend confirms and agrees with the study by Velázquez-Altamirano and Córdoba-Alcaráz (2020), called "Evaluation of the CIJ smoking cessation clinic", in which they found the need to include a greater number of patients from different geographical areas, socioeconomic levels and ages, not only to enrich individual experiences and diagnoses, but also to have greater coverage than in previous years.

The upper peaks observed in Figure 12 represent people with the most ingrained smoking habit and lower peaks represent those with the least ingrained habit, it is also observed that most of the patients are concentrated between the average values of 1.5 to 3, and only a small part exceeds the range of 4 to 6 points, a group that is identified as having a more ingrained level of addiction to tobacco.

Table 24, below, shows the exact value obtained for each participant, which as described in previous paragraphs, all participants were included in the study for being within the range established for viable candidates by equaling or exceeding the minimum consumption requested, for reasons of privacy and data protection, the researcher did not have access to personal and sensitive data of patients such as name, age, sex, among others, simply distinguishing each patient by the identifier assigned by the therapists.

Table24 *Averages of consumption according to the "Motives for Consumption" questionnaire obtained for each patient.*

	Average Obtained
Patient 1	2
Patient 2	2.05714286
Patient 3	2.14285714
Patient 4	1.62857143
Patient 5	1.8
Patient 6	1.74285714
Patient 7	2.22857143
Patient 8	1.97142857
Patient 9	1.91428571
Patient 10	1.82857143
Patient 11	2.14285714
Patient 12	2.77142857
Patient 13	2.8
Patient 14	2.71428571
Patient 15	4.02857143
Patient 16	3.8
Patient 17	3.31428571

Patient 18	3.51428571
Patient 19	2.94285714
Patient 20	2.65714286
Patient 21	2.31428571
Patient 22	2.85714286
Patient 23	2.71428571
Patient 24	2.4
Patient 25	1.85714286
Patient 26	2.65714286
Patient 27	2.4
Patient 28	2.85714286
Patient 29	2.57142857
Patient 30	2.62857143
Patient 31	2
Patient 32	2.02857143
Patient 33	2.17142857
Patient 34	1.8
Patient 35	2.11428571
Patient 36	1.62857143
Patient 37	2.22857143
Patient 38	2.11428571
Patient 39	1.82857143
Patient 40	2.08571429
Patient 41	2.25714286
Patient 42	2.62857143

Patient 43	2.74285714
Patient 44	2.6
Patient 45	3.65714286
Patient 46	3.6
Patient 47	3.08571429
Patient 48	3.4
Patient 49	2.88571429
Patient 50	2.51428571
Patient 51	2.48571429
Patient 52	2.82857143
Patient 53	2.28571429
Patient 54	2.48571429
Patient 55	1.94285714
Patient 56	2.57142857
Patient 57	2.37142857
Patient 58	2.54285714
Patient 59	2.74285714
Patient 60	2.54285714

Note: Prepared by the authors

Figure 13 takes as an example the answers to the first item in which it can be seen that the options "frequently" and "always" had no impact on the patients, in other words, they never selected those answers for this item.

Figure13 *Distribution of responses to the item "I feel that smoking gives me security".*

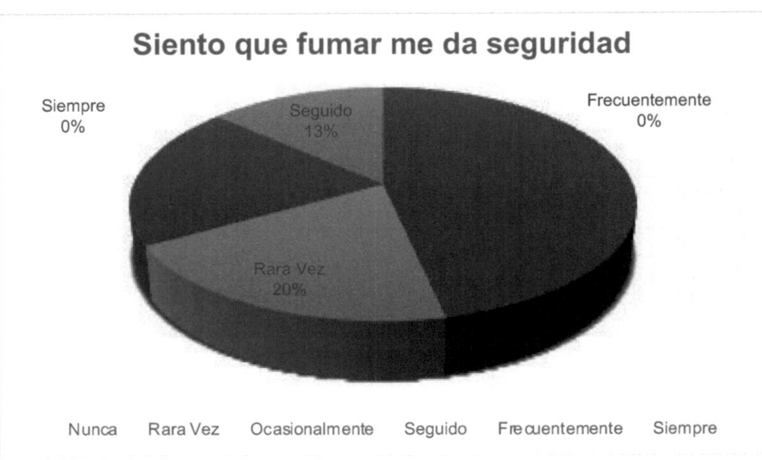

Note: Own elaboration (Velázquez, 2022).

The concurrence in the selection of answers by patients is expressed in Figure 13, obtaining a lower number of selections the options "Frequently" and "Always", this item is categorized within the psychosocial area by Velázquez-Altamirano and Córdova-Alcaráz (2020), in addition, the item and the others included in the questionnaire "Motives for tobacco use" has been reviewed and endorsed by the author Chavez Vizuet (2016), this instrument that has been used in smoking cessation treatment of Juvenile Integration Centers throughout the country.

Figure 14 shows the total number of options selected by the patients in the questionnaire. It can be highlighted that the options "Never" and "Rarely" are the ones that obtained the highest number of selections, that is, for all the questions the option "Rarely" was selected 584 times by the patients.

Figure14 *Total, responses selected by patients in the questionnaire "Motives for tobacco use".*

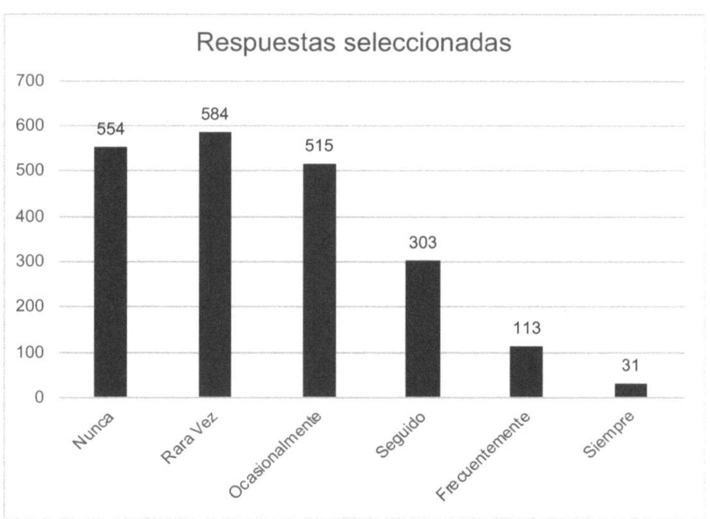

Note: Own elaboration (Velázquez, 2022).

Figure 15 shows the percentage of options selected by the patients in the questionnaire. It can be highlighted that the options "Never" and "Rarely" are the ones that obtained the highest percentage of selection, that is, for all the questions the option "Rarely" represents 28% in terms of selection by the patients.

Figure15 *Percentage, responses selected by patients in the questionnaire "Motives for tobacco use".*

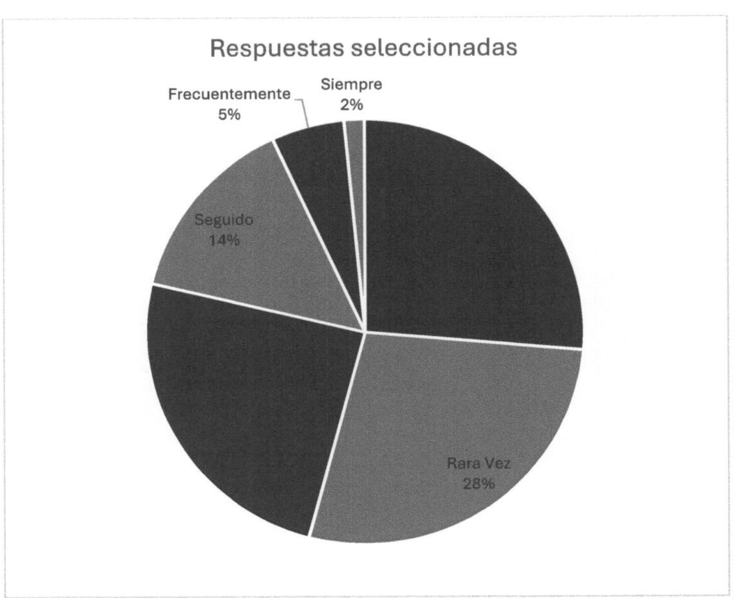

Note: Own elaboration (Velázquez, 2022).

Figure 16 shows the sum per question in the "Motives for consumption" questionnaire, which shows the three most common or frequent motives selected by the patients, with the result that questions 6, 16 and 26 were those to which the patients gave the highest scores, these questions correspond to the following statements:

6. I still feel the need for a cigarette, with a total of 189 points.

26. I smoke more when I am tense, with a total of 181 points.

16. In monotonous or boring jobs I smoke more, with a sum of 180.

Figure16 *Highest scoring items determining the three most frequent reasons for tobacco use in patients*

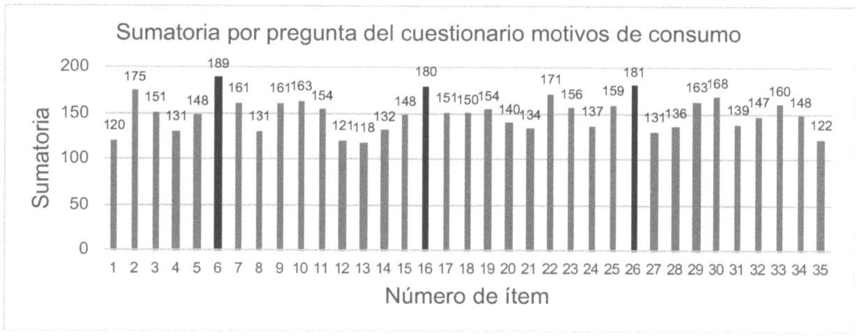

Note: Own elaboration (Velázquez, 2022).

The identification of the main reasons for tobacco use is essential for therapists, as stated by Velázquez-Altamirano and Córdoba-Alcaráz (2020), which affirms that this is intended to meet the objectives established in the program to successfully complete the treatment, and thus reduce or eliminate tobacco use, Londoño Pérez et al, (2021), also considered including this type of items in their studies because these questions can be taken as indicators to evaluate these aspects and thus strengthen the evidence in the precise classification of the type and reasons for consumption, in their study they did not include this type of questions, but still consider them important.

4.4.2 Results on Bot Usability

The usability of an application is the level at which a product can be used by end users to achieve objectives with effectiveness, efficiency and satisfaction in a given context of use (González and Reboredo, 2019) , for this purpose, the usability of the application was evaluated using the corresponding questionnaire described in Annex 2, this evaluation was performed more or less halfway through the treatment of the first group, the evaluation for the second group was performed at the end of the treatment, the obtaining of results was divided into 2 parts starting with the first group of 28 people in the month of February 2018, the second group started activities at the end of the month of April 2018 with a total of 32 people.

Once the questionnaire was applied, the results obtained were grouped according to the type of question used, which are grouped into three areas: performance, functionality and information. For the performance section (questions 1, 4, 12 and 13) the following results were obtained:

Question 1, The installation of the Telegram program, 100% of the respondents answered between the ranges Excellent, Very good and Good, so it is concluded that all respondents are familiar with the installation of applications on their mobile devices, the total percentages and the frequency of each range are expressed in Table 25.

Table25 *Descriptive statistics and frequencies related to question 1 of the questionnaire "Bot Usability".*

Installation					
		Frequency	Percentage	Valid percentage	Cumulative percentage
Valid	Good	15	25.0	25.0	25.0
	Very good	26	43.3	43.3	68.3
	Excellent	19	31.7	31.7	100.0
	Total	60	100.0	100.0	

Note: Prepared by the authors

Question 4, the response time, 58% of respondents decided to select the Excellent option, also 18.33% responded that the response time was Very good, this determines that the speed to receive messages as to open the application was agile, the Telegram application is considered lighter than other mobile messaging applications according to the website geeknetic , the total percentages and frequency of each range are expressed in Table 26.

Table26 *Descriptive statistics and frequencies related to question 4 of the questionnaire "Bot Usability".*

Response Time					
		Frequency	Percentage	Valid percentage	Cumulative percentage
Valid	Good	14	23.3	23.3	23.3
	Very good	11	18.3	18.3	41.7
	Excellent	35	58.3	58.3	100.0
	Total	60	100.0	100.0	

Note: Prepared by the authors.

Question 12, In the case of data consumption (internet), only 18% responded that consumption was Fair to Poor, while 81.67% expressed that data consumption was Good to Excellent, which translates into low data consumption.

Table27 *Descriptive statistics and frequencies related to question 12 of the questionnaire "Bot Usability".*

Data Consumption					
		Frequency	Percentage	Valid percentage	Cumulative percentage
Valid	Mala	1	1.7	1.7	1.7
	Regular	10	16.7	16.7	18.3
	Good	8	13.3	13.3	31.7
	Very good	23	38.3	38.3	70.0
	Excellent	18	30.0	30.0	100.0

	Total	60	100.0	100.0	

Note: Prepared by the authors.

Question 13, the overall execution speed of the Bot, this question is directly linked to question 4, in this question the respondents answered from Good to Excellent in 88.33%, the total percentages and the frequency of each range are expressed in table 28.

Table28 *Descriptive statistics and frequencies related to question 13 of the questionnaire "Bot Usability".*

		Execution			
		Frequency	Percentage	Valid percentage	Cumulative percentage
Valid	Regular	7	11.7	11.7	11.7
	Good	6	10.0	10.0	21.7
	Very good	25	41.7	41.7	63.3
	Excellent	22	36.7	36.7	100.0
	Total	60	100.0	100.0	

Note: Prepared by the authors.

The previous items included in the aspect denominated in the present study as "Performance", are assimilated to the dimension "Satisfaction" presented by Pérez Peña and Ramos Jurado (2021), where they obtained a very low qualification, with a low level of 56.67% of the respondents and a moderate level of 43.33% of the same, in a population of 30 people, with which they conclude that satisfaction is low in terms of customer service, contrary to the present study where a level of acceptance of over 80% was obtained.

For the functionality section, questions 2, 3, 5, 6, 7, 8, 9 were considered, with an average of 93% of positive responses among all the items and only 6.3% in those considered negativethe results of each of them are detailed below:

Question 2, Contact search, 91.67% of the respondents determined that the search for the contact "Quit Smoking" in Telegram was "Excellent", so it is determined that the search and aggregation of the contact was quick and easy, the total percentages and the frequency of each range are expressed in Table 29.

Table29 *Descriptive statistics and frequencies related to question 2 of the questionnaire "Bot Usability".*

Search Contact					
		Frequency	Percentage	Valid percentage	Cumulative percentage
Valid	Good	3	5.0	5.0	5.0
	Very good	2	3.3	3.3	8.3
	Excellent	55	91.7	91.7	100.0
	Total	60	100.0	100.0	

Note: Prepared by the authors.

Question 3, Bot help, the Bot as such does not provide a help system for patients or have a database on key terms or frequently asked questions, it simply provides a dictionary of the commands that can be used within the Bot, therefore only 50% of the people who answered this question rated as "Excellent" the dictionary shown when selecting "Help", the total percentages and the frequency of each rank are expressed in Table 30.

Table30 *Descriptive statistics and frequencies related to question 3 of the questionnaire "Bot Usability".*

Help

		Frequency	Percentage	Valid percentage	Cumulative percentage
Valid	Mala	3	5.0	5.0	5.0
	Regular	8	13.3	13.3	18.3
	Good	2	3.3	3.3	21.7
	Very good	17	28.3	28.3	50.0
	Excellent	30	50.0	50.0	100.0
	Total	60	100.0	100.0	

Note: Prepared by the authors.

Questions 5 and 6, clarity of the available options and clarity of the answers received, more than 60% of the respondents decided as "Excellent" for both questions, the total percentages and the frequency of each rank are shown in table 31 and 32 respectively.

Table31 *Descriptive statistics and frequencies related to question 5 of the questionnaire "Bot Usability".*

Available Options					
		Frequency	Percentage	Valid percentage	Cumulative percentage
Valid	Regular	1	1.7	1.7	1.7
	Good	9	15.0	15.0	16.7
	Very good	11	18.3	18.3	35.0
	Excellent	39	65.0	65.0	100.0
	Total	60	100.0	100.0	

Note: Prepared by the authors.

Table32 *Descriptive statistics and frequencies related to question 6 of the questionnaire "Bot Usability".*

Responses Received					
		Frequency	Percentage	Valid percentage	Cumulative percentage
Valid	Mala	1	1.7	1.7	1.7
	Regular	1	1.7	1.7	3.3
	Good	2	3.3	3.3	6.7
	Very good	19	31.7	31.7	38.3

	Excellent	37	61.7	61.7	100.0
	Total	60	100.0	100.0	

Note: Prepared by the authors.

Question 7, ease of use of the Bot, 76.67% of the patients determined that the handling of the Bot was "Good" to "Excellent", and 21.67% selected the option of "Regular", the rest determined that it was bad and therefore considered difficult for a certain segment of the population evaluated (1.67%), the total percentages and the frequency of each range are shown in Table 33.

Table33 *Descriptive statistics and frequencies related to question 7 of the questionnaire "Bot Usability".*

		Facility			
		Frequency	Percentage	Valid percentage	Cumulative percentage
Valid	Mala	1	1.7	1.7	1.7
	Regular	13	21.7	21.7	23.3
	Good	1	1.7	1.7	25.0
	Very good	14	23.3	23.3	48.3
	Excellent	31	51.7	51.7	100.0
	Total	60	100.0	100.0	

Note: Prepared by the authors.

Questions 8 and 9, the graphical interface of the Bot and The type and size of the text, this element received in general good acceptance: 98.33% and 75% respectively, due to the interface presents a design equal to that of the chats found in the most popular instant messaging applications, the total percentages and the frequency of each range are expressed in Table 34 and 35 respectively.

Table34 *Descriptive statistics and frequencies related to question 8 of the questionnaire "Bot Usability".*

		Interface			
		Frequency	Percentage	Valid percentage	Cumulative percentage
Valid	Regular	1	1.7	1.7	1.7
	Very good	18	30.0	30.0	31.7
	Excellent	41	68.3	68.3	100.0
	Total	60	100.0	100.0	

Note: Prepared by the authors.

Table35 *Descriptive statistics and frequencies related to question 9 of the questionnaire "Bot Usability".*

		Text			
		Frequency	Percentage	Valid percentage	Cumulative percentage
Valid	Regular	1	1.7	1.7	1.7
	Good	14	23.3	23.3	25.0
	Very good	17	28.3	28.3	53.3
	Excellent	28	46.7	46.7	100.0
	Total	60	100.0	100.0	

Note: Prepared by the authors.

In the previous items included in the aspect denominated in the present study as "Functionality", they obtained as results on average a very good acceptance by the respondents, similar to what was reported by Labra Chino and Quispe Poma (2022), where through the analysis of the results obtained through their instruments, they demonstrated that the chat Bot is efficient in response time with an average of 0.8 milliseconds in real time responses as well as in the overall performance of the chat Bot, determining that it is flexible and optimal, given that the chat Bot's capacity is adaptable.

The results of both studies reflect Bot's flexibility in execution due to the low resources consumed on the devices and the high response and availability rates supported by their respective vendors' technology platforms.

For the information section, questions 10 and 11 were considered, with an average of 100% of responses considered as positive for the two items involved, as follows is the result of each one of them:

Question 10, tips received (subject matter), 100.00% of the respondents determined that the tips received through Bot and the subject matter related to smoking were between "Excellent", "Very good" and Good", so it is determined that these actions are accepted by most patients, the total percentages and the frequency of each range are shown in Table 36.

Table36 *Descriptive statistics and frequencies related to question 10 of the questionnaire "Bot Usability".*

Tips					
		Frequency	Percentage	Valid percentage	Cumulative percentage
Valid	Good	15	25.0	25.0	25.0
	Very good	24	40.0	40.0	65.0
	Excellent	21	35.0	35.0	100.0
	Total	60	100.0	100.0	

Note: Prepared by the authors.

Question 11, periodicity of tips received, 93.4% of respondents determined that the frequency tips received, i.e. the time between one tip and another or more specifically the tips received per day was between "Excellent", "Very good" and "Good", only 6.7% determined that this action is "Regular", so it is determined that the frequency of tips received is very well accepted, the total percentages and the frequency of each range are expressed in Table 37.

Table37 *Descriptive statistics and frequencies related to question 10 of the questionnaire "Bot Usability".*

		Periodicity Tips			
		Frequency	Percentage	Valid percentage	Cumulative percentage
Valid	Regular	4	6.7	6.7	6.7
	Good	6	10.0	10.0	16.7
	Very good	25	41.7	41.7	58.3
	Excellent	25	41.7	41.7	100.0
	Total	60	100.0	100.0	

Note: Prepared by the authors.

In the previous items included in the aspect denominated in the present study as "Information", the results showed a very high level of acceptance by valuing the total selection of answers considered as positive in 100% by all respondents, similar to what was obtained in the study presented by Villegas-Ch et al, (2020), in which they determined that the right information at the right time is knowledge and that in addition to the basic tasks programmed in a Bot, it is also important to include additional information within its design, to reinforce the knowledge and retention of the information in the users.

The concurrence in the selection of responses by patients is expressed in Figure 17, with the highest number of selections being obtained for the options "Excellent" and "Very good".

Figure17 *Total, responses selected by patients in the "Bot Usability" questionnaire.*

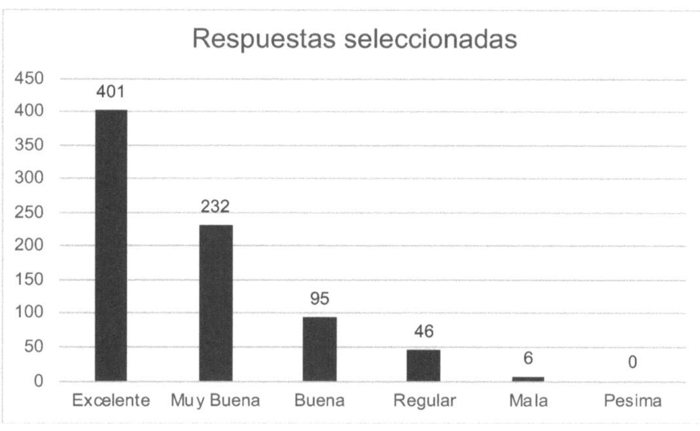

Note: Own elaboration (Velázquez, 2022).

Taking into consideration the sum of the selection of the options considered as positive; "Excellent", "Very Good" and "Good", it can be considered that the Bot, in general terms, was considered as "Highly usable" for 93% of the patients.

Figure 18 shows the percentage of options selected by the patients in the questionnaire in general for all the questions. It can be highlighted that the options "Excellent" and "Very Good" are the ones that obtained the highest percentage of selection, with 51% and 31%, respectively.

Figure18 *Percentage, answers selected by patients in the questionnaire "Bot Usability".*

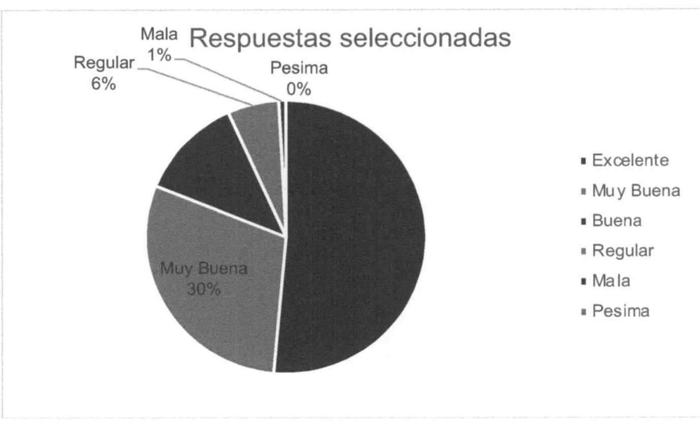

Note: Own elaboration (Velázquez, 2022).

Table 38 below shows the selection trends of the present questionnaire, referring to the averages per patient regarding their opinion on the usability of the Bot, clearly showing a dominant trend in the general consideration of each individual represented by the parameters "Very good" and "Excellent".

Table38 *Selection trend "Bot Usability".*

		Frequency	Percentage	Valid percentage	Cumulative percentage
Valid	Very Good	13	21.7	21.7	21.7
	Excellent	47	78.3	78.3	100.0
	Total	60	100.0	100.0	

Note: Prepared by the authors.

Figure 19 graphically represents the dominant selections in the "Bot Usability" questionnaire, with the "Very good" and "Excellent" options standing out predominantly.

Figure19 *Graph on the selection trend "Bot Usability".*

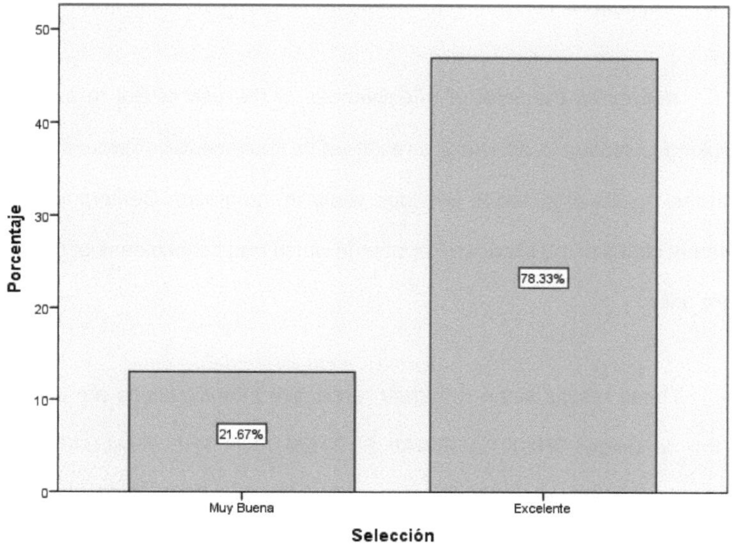

Note: Own elaboration (Velázquez, 2022).

4.4.3 Results on treatment efficacy

This section is considered the core of the research because it is directly related to the hypothesis of this document associated with the management processes of the same, it is due to the satisfactory completion of the treatment by the patients and to interpret its results the questionnaire corresponding to the effectiveness of the treatment was used for each patient, with data on the response to treatment using the Bot.

This instrument is answered by the therapist at the end of each treatment, always respecting the times established for each particular patient, and a web platform will be made available to the professionals to capture the results regarding the progress of each patient, thus speeding up the data collection process.

determine the level of effectiveness of the use of Bot in support of smoking cessation treatment, the results of its implementation were compared with the results obtained in previous years in Integration Centers located in different states of the Mexican Republic in which traditional means of treatment were used.

These results along with their respective interpretations are published in the document "INSTITUTIONAL SYSTEM FOR THE EVALUATION OF TREATMENT PROGRAMS EVALUATION OF THE CIJ STOP SMOKING CLINIC" the objective of this document is to "retrospectively evaluate the results of the Smoking Cessation Treatment Program applied to patients attending the care units of Juvenile Integration Centers (Velázquez-Altamirano and Córdova-Alcaráz, 2020) .

As a first comparative analysis, Figure 20 refers to Graph 1 of the document mentioned in the previous paragraph, which mentions "Of the patients who came to the last session, the majority consumed up to 10 cigarettes a day before treatment (41.3%), a little more than a third smoked between 11 and 20, a quarter consumed between 21 and 30 and a smaller proportion consumed more than 31 cigarettes daily (4.7%)", in the results obtained from the application of the "Treatment Effectiveness" questionnaire.

Therapists determined on average for question number 1 "Reduction of consumption in units per day" a 55% from Regular to Excellent according to their experience according to the reduction of consumption in each patient, so that more than half of the patients presented significant progress in terms of reduction of daily cigarette consumption.

Figure20 *Daily cigarette consumption, before and after treatment.*

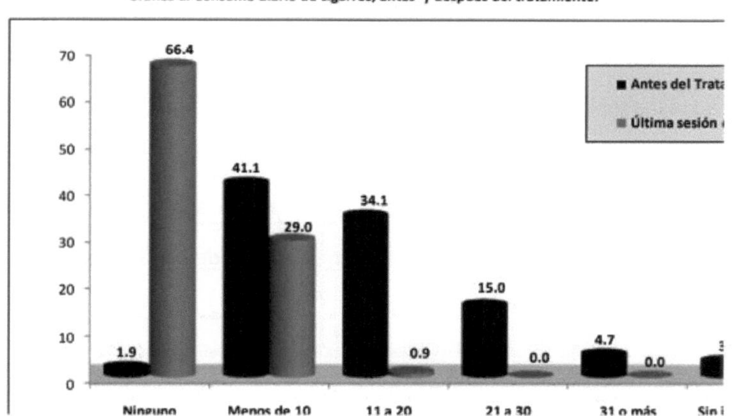

Note. Taken from (Velázquez-Altamirano and Córdova-Alvaráz, 2020, p.x. 10).

Question 9 "In general, what was the patient's progress with this treatment", also refers to the progress presented according to the therapist's opinion of each patient, in this question 56.67% of the patients had general progress throughout the treatment taking into account the use of Bot.

Comparing the results Tudor Sfetea et al. (2018), reports in his study as results the reduction of daily consumption by 53%, a figure slightly lower than that represented in the present study, on the other hand Lin et al. (2018), reports 59.6% in terms of the reduction of daily consumption per unit of tobacco, analyzing the three results, it can be argued that Bot, on average, is within the range of the results reported by other authors, although the studies are similar, in the present study other reagents were added to strengthen the analysis of the effectiveness of Bot in smoking cessation treatment.

Another series of questions were included in the questionnaire "effectiveness of the treatment" (2, 3, 4, 5, 6, 7 and 8), which are directly related to the topic mentioned in this section, in which in general an acceptable percentage of acceptance and management of the treatment was obtained, 59.29% of the answers to the previous questions on average were marked from Fair to Excellent, the results of each of them are detailed in Table 39.

Table39 *Results obtained from the Treatment Effectiveness Questionnaire*

Ask	Positive responses	Negative responses
The use of time dedicated to treatment was	65%	33.33%
Reduction of consumption time	61.67%	36.67%
Incident reduction	51.67%	46.60%
Increase in successful therapies	66.67%	30%
Zero therapy reduction	56.67%	41.67%

Reduction of logbook misplacement	60%	40%
Reduction of incorrect filling of logbooks	53.33%	46.67%

Source: Own elaboration

In this dimension, related to complementary efforts in the treatment against smoking, slightly positive percentages were obtained in each of the items unlike the mostly positive results obtained in the study by Tudor Sfetea et al. (2018), in which they state that their desire to continue using the application is 67% of people and that they would recommend the application to another person reaches up to 73%

For their part Alsharif and Philip (2015), report in their study, that the mobile application they designed obtained the following percentages of acceptance; progress graphs 90%, alerts and notifications 57.6%, informative videos, educational material, feedback from doctors and ex-smokers 88.4%, general help, helpline 77%, continue with the application at the end of treatment 73%.

Based on these results, it could be stated that the Telegram Bot used in the present study fulfilled the management and administration of smoking cessation treatment, i.e. support in administrative tasks and organization of information, lightening the work of therapists and health professionals as well as the control of records by the users, in terms of the results related to the effectiveness of the treatment it did not have a considerable impact associated with the reduction in consumption per day.

In relation to the concurrence in the selection of responses by the therapists associated with the progress and follow-up, with respect to the treatment against smoking, as shown in Figure 21, a greater number of selections were obtained in the options "Poor" and "fair".

Figure21 *Total, responses selected by the therapists in the questionnaire "Effectiveness in treatment".*

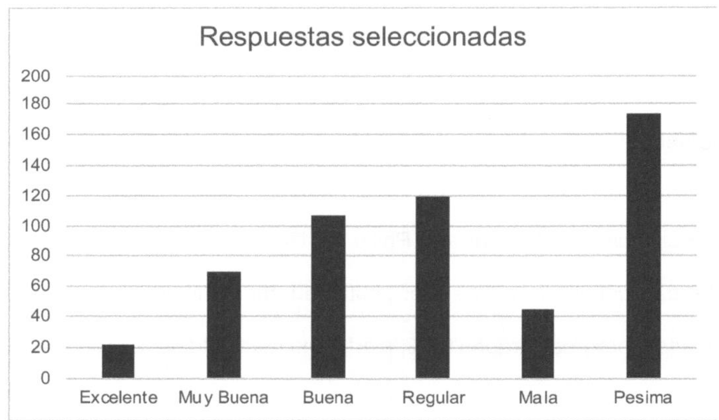

Note: Own elaboration (Velázquez, 2022).

Figure 22 shows the percentages obtained taking into account the sum of the selection of the options classified as positive; "Excellent", "Very Good" and "Good", it can be considered that the efficacy of the treatment in general terms was satisfactory for 37% of the patients, while for the remaining 63% it is considered that their progress was regular to null.

Figure22 *Percentage, answers selected by the therapists in the questionnaire "Effectiveness in treatment".*

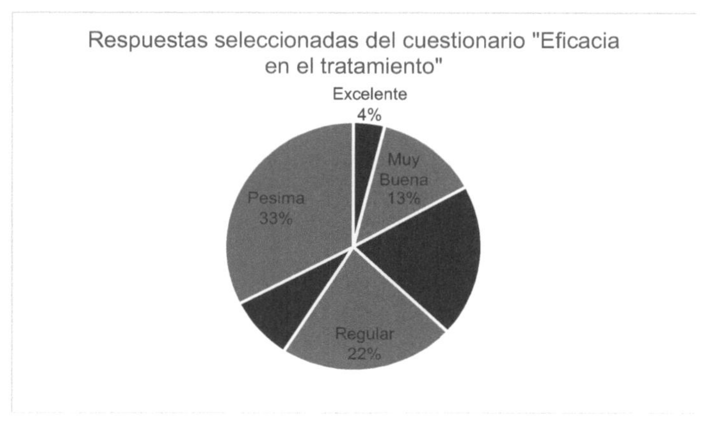

Note: Own elaboration (Velázquez, 2022).

Table 40 below shows the selection trends of the present questionnaire, referring to the selection averages per therapist in terms of their opinion on the effectiveness of the treatment in each of the patients, clearly showing a dominant trend in the general consideration of each therapist represented by the parameters of "Fair" in first place and "Good" in second place.

Table40 *Selection trend "Treatment efficacy".*

		Frequency	Percentage	Valid percentage	Cumulative percentage
Valid	Lousy	6	10.0	10.0	10.0
	Mala	5	8.3	8.3	18.3
	Regular	24	40.0	40.0	58.3
	Good	15	25.0	25.0	83.3
	Very good	9	15.0	15.0	98.3
	Excellent	1	1.7	1.7	100.0
	Total	60	100.0	100.0	

Note: Prepared by the authors.

Figure 23 graphically represents the dominant selections in the "Effectiveness in treatment" questionnaire, with the "Regular" option standing out predominantly.

Figure23 *Graph on the selection trend "Treatment Effectiveness".*

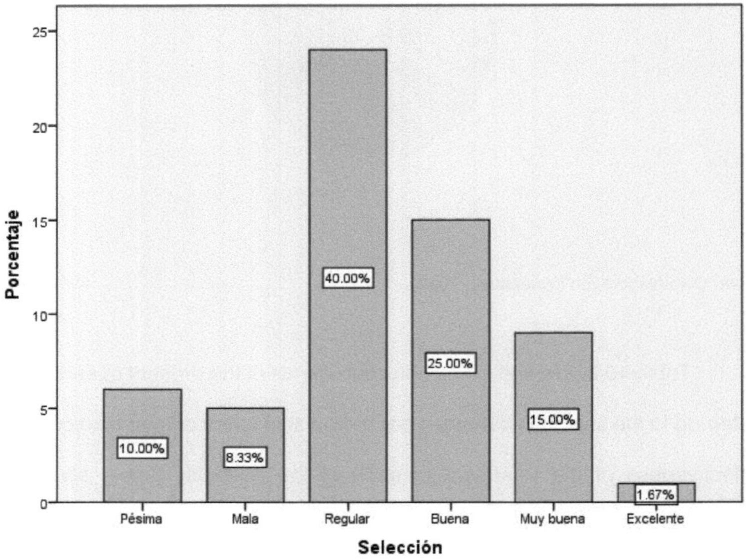

Note: Own elaboration (Velázquez, 2022).

4.4.4 Results on the usability of the Web platform

This section shows the results obtained according to the therapists' satisfaction with the website that allowed them to complement the use of the Bot in the management of smoking cessation treatment; this site allowed them to perform consultations by patient, history and records added by the patients.

This analysis took into account the achievement of task completion and user satisfaction while performing the tasks

If users are able to accomplish what they want to do within the site, do not take too much time to do it, perceive a certain ease in performing actions, usually do not make many mistakes and have a high degree of mastery of the application after using it, then the software product is considered usable.

These metrics were evaluated in the usability questionnaire of the web platform (see annex 4), by means of 10 questions which were divided into 2 sections: the first one referring to highlight positive issues of the site, and the second to mention possible negative issues found in the operation or appearance of the site, this questionnaire was applied to 5 people who were the only ones who used the site.

For the first section, the following items were used, obtaining the following results:

Question 1, I think I would like to visit this Web site frequently. 80% of respondents selected the options "I agree" and "I strongly agree", 20% selected the option "I somewhat agree", in this reactive refers to the personal intention to revisit the website according to the perception that they may have perceived in previous visits, so it is determined that the frequency in terms of returning to visit the page or not with very good acceptance, the total percentages and frequency of each range are expressed in Table 41.

Table41 *Descriptive statistics and frequencies related to question 1 of the questionnaire "Usability of the Web platform".*

Visit Web site					
		Frequency	Percentage	Valid percentage	Cumulative percentage
Valid	Somewhat in agreement	1	20.0	20.0	20.0
	Agreed	2	40.0	40.0	60.0
	I strongly agree	2	40.0	40.0	100.0
	Total	5	100.0	100.0	

Note: Prepared by the authors.

Question 5, I found the various possibilities of the Web site to be quite well integrated. Of the respondents 60% determined that "I agree" and 20% selected the option "I strongly agree" in the situation raised, adding up to a total of 80% in positive reactions, unlike other usability studies where the integrated options are evaluated where 71% of respondents consider that the structure of the page and the information contained in each section is sufficiently clear (Serrano Mascaraque, 2009) , with this level of percentage the author considers that his web page is usable because the users easily find the different integrated options through the complete interface of his website, therefore, the web platform that supports the Bot's work meets the expectations of the aforementioned author.

Table 42 shows that the frequency of returning to visit the page or not, is established with very good acceptance, in addition to the total percentages and the frequency of each range.

Table42 *Descriptive statistics and frequencies related to question 5 of the questionnaire "Usability of the Web platform".*

Integrated possibilities					
		Frequency	Percentage	Valid percentage	Cumulative percentage
Valid	Somewhat in agreement	1	20.0	20.0	20.0
	Agreed	3	60.0	60.0	80.0
	I strongly agree	1	20.0	20.0	100.0
	Total	5	100.0	100.0	

Note: Prepared by the authors.

Question 7, I imagine that most people would learn very quickly to use the Web site, this topic related to the perception of each respondent on how quickly they consider to learn to use the site, resulted in the following percentages, 40% selected "I strongly agree" and another 40% selected "I agree", totaling 80% of acceptance.

In contrast to the results obtained by Serrano Mascaraque (2009), where he found in a similar item that 42% of the participants consider that access to information and the speed to navigate the site is similar using different types of browsers, without finding differences or difficulties to do so.

Therefore, it is determined that the frequency regarding the speed of learning the website, is established with very good acceptance, the total percentages and the frequency of each range are expressed Table 43.

Table43 *Descriptive statistics and frequencies related to question 7 of the questionnaire "Usability of the Web platform".*

Learning speed					
		Frequency	Percentage	Valid percentage	Cumulative percentage

Valid	Somewhat in agreement	1	20.0	20.0	20.0
	Agreed	2	40.0	40.0	60.0
	I strongly agree	2	40.0	40.0	100.0
	Total	5	100.0	100.0	

Note: Prepared by the authors.

Question 9, I felt very confident in the handling of the Web site. 100% of the respondents agree at some level of conviction with the question posed, therefore, it is determined that the frequency regarding the speed of learning the website, is established with very good acceptance, the total percentages and the frequency of each range are expressed in Table 44.

Table44 *Descriptive statistics and frequencies related to question 9 of the questionnaire "Usability of the Web platform".*

Confidence in website management					
		Frequency	Percentage	Valid percentage	Cumulative percentage
Valid	Somewhat in agreement	1	20.0	20.0	20.0
	Agreed	3	60.0	60.0	80.0
	I strongly agree	1	20.0	20.0	100.0
	Total	5	100.0	100.0	

Note: Prepared by the authors.

For the second part related to possible negative points found in the platform in general, following results were obtained on average for questions 2, 3, 4, 6, 8 and 10:

- 60% selected "I do not agree at all".

- 20% selected "I do not agree".
- 16.67% "Somewhat disagree".
- 3.33% "Somewhat agree".

Question 2, I found the website not complex, refers to the ease of use and navigation perceived by the user, in which the sum of the percentages obtained through the selection of the users was 100%, in contrast to the results obtained by Serrano Mascaraque (2009), determined that there are elements in their web page that make navigation difficult or the navigation system is difficult to use, finding that 30% of the respondents consider that there are elements that make navigation difficult, either because of the poor distribution of elements, the advertising of the site or because some other elements are superimposed on the main ones.

Therefore, it is determined that the frequency regarding the perception of complexity within the website, is established with very good acceptance, the total percentages and the frequency of each range are expressed in table 45.

Table45 *Descriptive statistics and frequencies related to question 2 of the questionnaire "Usability of the Web platform".*

		Website not complex			
		Frequency	Percentage	Valid percentage	Cumulative percentage
Valid	Agreed	1	20.0	20.0	20.0
	I strongly agree	4	80.0	80.0	100.0
	Total	5	100.0	100.0	

Note: Prepared by the authors.

Question 3, it is relatively easy to use the Web site, refers to the ease of use and navigation perceived by the user, as the previous question relates to the perception of ease of use perceived by users, in which the sum of the percentages obtained through the selection of users was 100%, in agreement with the results obtained by Serrano Mascaraque (2009), 64% of the respondents in his study determined that this percentage of users reach most of the options with two clicks, he explains that this may be due to previous experience or easy integration of skills related to the tasks of your website.

Therefore, it is determined that the frequency in terms of perceived ease of using the website is established with very good acceptance, the total percentages and the frequency of each range are expressed in Table 46.

Table46 *Descriptive statistics and frequencies related to question 3 of the questionnaire "Usability of the Web platform".*

Ease of use of the web site					
		Frequency	Percentage	Valid percentage	Cumulative percentage
Valid	Agreed	2	40.0	40.0	40.0
	I strongly agree	3	60.0	60.0	100.0
	Total	5	100.0	100.0	

Note: Prepared by the authors.

Question 4, I do not need the support of an expert to browse the Web site, refers to the ease of use and navigation without the need for prior training, like the previous two questions, it is related to the perception of ease of use perceived by users, in this item, the sum of the percentages obtained through the selection of users was 100%.

Therefore, it is determined that the frequency regarding the perceived ease of using the website is established with very good acceptance, the total percentages and the frequency of each range are expressed in Table 47.

Table47 *Descriptive statistics and frequencies related to question 4 of the questionnaire "Usability of the Web platform".*

		Training not required			
		Frequency	Percentage	Valid percentage	Cumulative percentage
Valid	Agreed	1	20.0	20.0	20.0
	I strongly agree	4	80.0	80.0	100.0
	Total	5	100.0	100.0	

Note: Prepared by the authors.

Question 6, is there consistency in the Web site, refers to the users' perception of the congruence and concordance between all its elements, in this item, the sum of the percentages obtained through the users' selection of the selected options was 80%.

Therefore, it is determined that the frequency regarding the perception of consistency in the website is established with good acceptance, the total percentages and the frequency of each range are expressed in table 48.

Table48 *Descriptive statistics and frequencies related to question 6 of the questionnaire "Usability of the Web platform".*

		Website consistency			
		Frequency	Percentage	Valid percentage	Cumulative percentage
Valid	Somewhat in agreement	1	20.0	20.0	20.0
	Agreed	1	20.0	20.0	40.0
	I strongly agree	3	60.0	60.0	100.0

	Total	5	100.0	100.0	

Note: Own elaboration

Question 8, I found the Web site visually attractive, refers to the users' perception of the condition of whether a Web site is pleasing to the eye, in this item, the sum of the percentages obtained through the users' selection of the selected options was 60%.

According to the results of Serrano Mascaraque (2009), the content of web pages should be displayed correctly and be pleasing to the eye, as this determines the permanence of the user on the website. In their study, 34% agreed with this characteristic

Therefore, it is determined that the frequency regarding the perception of visual attractiveness in the web page of the present study is established with good acceptance, the total percentages and the frequency of each range are expressed in Table 49.

Table49 *Descriptive statistics and frequencies related to question 8 of the questionnaire "Usability of the Web platform".*

		Visual appeal			
		Frequency	Percentage	Valid percentage	Cumulative percentage
Valid	Somewhat disagree	1	20.0	20.0	20.0
	Somewhat in agreement	1	20.0	20.0	40.0
	I strongly agree	3	60.0	60.0	100.0
	Total	5	100.0	100.0	

Note: Prepared by the authors.

Question 10, I don't need to learn many things before using the Web site, refers to the users' perception of the condition of whether the Web site can be manipulated by people without previous computer knowledge, in this item, the sum of the percentages obtained through the users' selection of the selected options was 40%.

According to the results of Serrano Mascaraque (2009), the content of web pages should be displayed correctly and be pleasing to the eye, as this determines the permanence of the user on the website. In their study, 34% agreed with this characteristic

Therefore, it is determined that the frequency regarding the perceived need for training in the use of the web page of the present study is established with regular acceptance, the total percentages and the frequency of each range are expressed in Table 50.

Table50 *Descriptive statistics and frequencies related to question 10 of the questionnaire "Usability of the Web platform".*

I don't need to learn a lot of things before I can use the .					
		Frequency	Percentage	Valid percentage	Cumulative percentage
Valid	Somewhat in agreement	3	60.0	60.0	60.0
	Agreed	1	20.0	20.0	80.0
	I strongly agree	1	20.0	20.0	100.0
	Total	5	100.0	100.0	

Note: Prepared by the authors.

In summary, the level of acceptance by users towards the web portal is 87.66%, following some metrics to evaluate the usability of a site, which mention that, having a high degree of acceptance, influence user satisfaction, the correct and efficient performance of the work is what determines the degree of acceptance of a product and therefore its usability

The concurrence in the selection of responses by patients is expressed in Figure 24, with the highest number of selections being obtained for the options "Excellent" and "Very good".

Figure24 *Total, responses selected by patients in the "Bot Usability" questionnaire.*

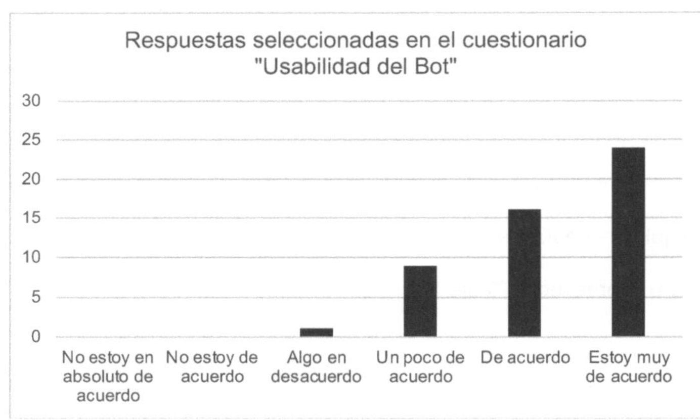

Note: Own elaboration (Velázquez, 2022).

Taking into consideration the sum of the selection of the options considered as positive; "I strongly agree", "I agree" and "I somewhat agree", we can consider that the level of usability of the Web platform in general terms was considered "High" to "Very high" for 98% of the patients, while for the remaining 2% it was considered fair in the dimensions evaluated, as shown in Figure 25.

Figure25 *Percentage, answers selected by the therapists in the questionnaire "Usability of the Web platform".*

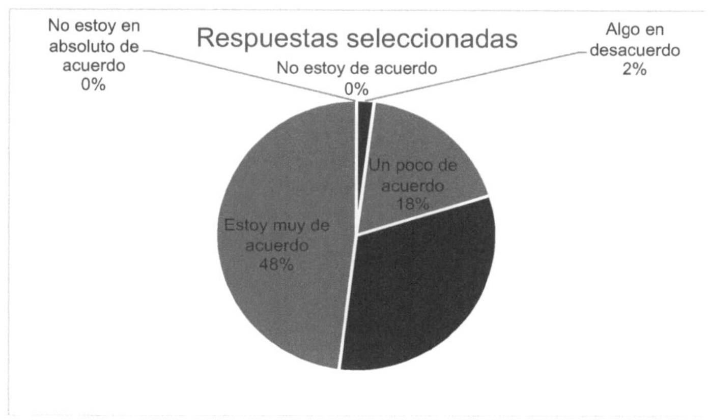

Note: Own elaboration (Velázquez, 2022).

Table 51 shows the selection trends of the present questionnaire, referring to the selection averages per therapist regarding their opinion on the usability of the Web platform, clearly showing a dominant trend on the general consideration of each therapist represented by the parameters "I agree" in second place and "I strongly agree" in first place, other selections are not shown because they have a percentage of zero.

Table51 *Trend of selection of the questionnaire "Usability of the web platform".*

		Frequency	Percentage	Valid percentage	Cumulative percentage
Valid	Agreed	1	20.0	20.0	20.0
	I strongly agree	4	80.0	80.0	100.0

Total	5	100.0	100.0

Note: Prepared by the authors.

Figure 26 graphically represents the dominant selections in the "Effectiveness in treatment" questionnaire, with the "I strongly agree" option standing out predominantly.

Figure26 *Graph on the selection trend "Usability of the Web platform".*

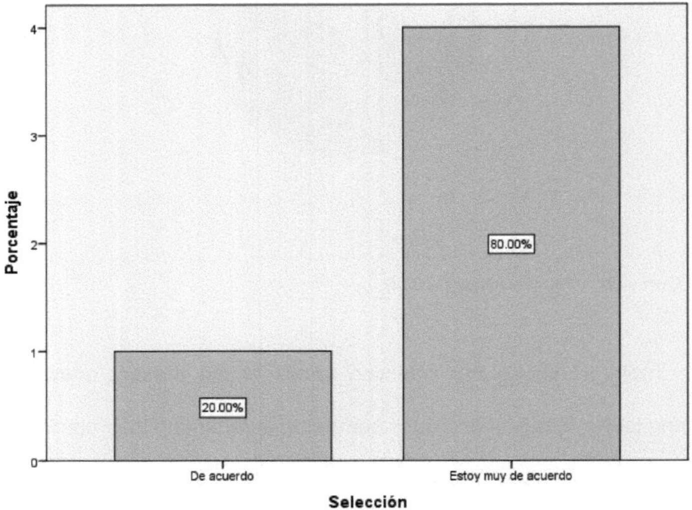

Note: Own elaboration (Velázquez, 2022).

4.4.5 Parametric test results

4.4.5.1 Pearson

Pearson's correlation test was used to determine whether "There is" or "There is not" a relationship between the two variables involved in this study, this relationship is also known as the level of significance, in this test the Pearson "r" Coefficient is applied to determine the degree of correlation.

If the significance level (Sig.) is equal to or less than 0.05, it is considered to have a confidence level of 95%, therefore, the degree of the correlation coefficient "r" is considered "significant".

If the significance level (Sig.) is less than or equal to 0.01, it is considered to have a 99% confidence level, therefore, the degree of the correlation coefficient "r" is considered "highly significant".

In this case it is observed that the Pearson correlation coefficient is 0.106, so it can be stated that in the study area there is a "very low positive correlation", between the dimension "Reduction of tobacco consumption" and "Bot for mobile devices", because the significance value is 0.418, which is below the required 0.05, the details of these values are expressed in Table 52 and 53.

Table52 *Descriptive statistics among the variables involved in the project*

Descriptive statistics			
	Media	Standard deviation	N
Consumption Reduction	26.0167	9.36183	60
Bot	68.2667	4.65729	60

Note: Prepared by the authors.

Table53 *Total, responses selected by patients in the questionnaire "Bot Usability".*

Correlations			
		Consumption Reduction	Bot
Consumption Reduction	Pearson correlation	1	.106
	Sig. (bilateral)		.418
	N	60	60
Bot	Pearson correlation	.106	1
	Sig. (bilateral)	.418	
	N	60	60

Note: Prepared by the authors.

4.4.5.2 Linear Regression

The linear regression function is a mathematical function that shows the causal relationship that exists between variables, through the relationship: y = a + bx; where "y" is the dependent variable and "x" the independent variable, the coefficient of the linear correlation allows determining the degree of association that exists in the dependence relationships of the variables considered, in other words, it measures the intensity between the variables being considered in the analysis, as shown in Figure 27, the corresponding scatter plot of the variables in question of this study.

Figure27 *Scatter plot of the variables "Consumption reduction" and "Bot for mobile devices".*

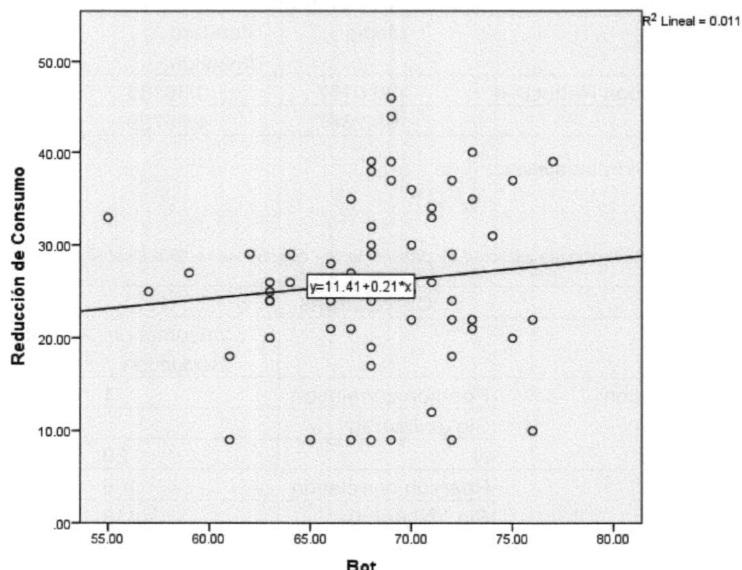

Note: Own elaboration, graph of the linear regression function and central adjustment line (Velázquez, 2022).

After performing the pertinent calculations, a correlation value between the two variables of 0.106 was obtained, considered by Mason and Lind (1995) as a "weak positive correlation"; a positive R value indicates a positive correlation, in which the values of the two variables tend to increase together. Following the interpretation of Table 54, 1.1% is considered as the reduction in consumption, the remaining 98.9% is attributed to variables not considered in this study, the details of these values are expressed in the table mentioned above.

Table54 *R Correlation calculated with the SPSS program*

Summary of the model				
Model	R	R square	Adjusted R-squared	Standard error of the estimate
1	.106[a]	.011	-.006	9.38858
a. Predictors: (Constant), Bot				

Note: Prepared by the authors.

The linear regression function is expressed with the values obtained from the unstandardized coefficients (B) in Table 55:

y = 11.415 + 0.214x

where y =0.011

The graph of the above function and its central adjustment line are shown in Figure 27.

Table55 *R correlation and coefficients obtained from SPSS program*

Coefficients[a]

Model		Unstandardized coefficients		Standardized coefficients	t	Sig.
		B	Standard error	Beta		
1	(Constant)	11.415	17.957		.636	.528
	Bot	.214	.262	.106	.815	.418

a. Dependent variable: Consumption Reduction.

Note. Own elaboration.

CHAPTER V CONCLUSIONS

5.1 Conclusions

In this thesis, a support tool called Bot for smoking cessation treatment at the Zacatecas Juvenile Integration Center was designed, implemented and experimentally validated. The tool approaches the design of a new application by generating an implementation and management of the treatment, which should be refined and updated according to the new standards of the Center to obtain a fully functional system that meets all the requirements of the users.

The objective of the support tool called Bot was to support in smoking cessation treatment and in the management of patient logs and information for treating physicians, reducing the necessary time and facilitating the development of such activities, aligned with the objectives of this paper; "To determine the impact of the use of a Telegram Bot for the management of patients under treatment for smoking cessation", it can be concluded that the objectives were partially achieved due to the low correlation between the study variables, i.e. the minimal impact that the Bot had on the treatment for smoking cessation, which, although positive, was not significant.

Therefore, the general hypothesis "The use of a Telegram Bot increases the impact on the management of smoking cessation treatment" is rejected, and the alternative hypothesis "The use of a Telegram Bot partially increases the impact on the management of smoking cessation treatment" is accepted."Although the impact of the tool on treatment was not as expected, i.e. the reduction of tobacco consumption compared to traditional methods did not have a significant improvement, it did have an impact on treatment management, lightening the workload of health professionals and safeguarding the records generated by each patient as part of their treatment.

With respect to the results obtained, although a favorable response was obtained, it is possible that it could be improved to provide better care and cover more niches of opportunity present in this institution, unfortunately the lack of budget and the constant rotation of personnel make follow-up difficult and delay its evolution, given that the treating physicians gave an outstanding evaluation of the application, it can be concluded that the objective of improving the management processes of patients undergoing treatment for smoking was achieved.

In addition to the above, the tool incorporated various technologies and programming languages in its implementation, all of which are free to use and documented with a view to possible future adaptations, evaluations, assessments and updates.

In 2018 when this tool was implemented there were still not many similar applications so users were not yet accustomed to using this type of interaction through instant messaging and even less with a semi-automatic response system.

The minimal increase in the effectiveness of treatment using this tool according to the instruments applied was largely due to the resistance used by patients and therapists to the use of new technologies.

To date (2023) there are multitudes of Bots in practically all the help systems of the large Web sites of commercial chains, banking institutions and an infinite number of businesses, this simplifies the work of the support area and channels the help more efficiently, which is why this application is considered a spearhead in this type of technology.

The proposed tool allows the incorporation of different emerging technologies for implementation, application and execution, taking advantage of their benefits, but above all their capabilities, not to mention that they are all free tools with great support in the network.

In short, I he possibility of including different multiplatform software architectures, and different programming languages, allows programming the proposed tool for other lines of the institution.

5.2 Recommendations

It is recommended to implement the use of the Bot in other sites with a larger number of participants in order to enrich the sample, with the objective of obtaining greater precision in the results according to the instruments used, with the purpose of evaluating the performance of the Bot in comparison to traditional treatment methods.

2. To enrich the instruments in order to cover all the dimensions involved and thus have an updated evaluation method that is more in line with new trends in treatment and user habits.

3. It is recommended to delve into documentation on smoking to enrich the Bot's knowledge base and thus provide a better user experience by having a more natural, broad and enriched communication.

4. Extend the number of tests to Bot considering different execution scenarios in order to have a more accurate parameter about its performance, data consumption and overall performance.

5. It is recommended to consult updated bibliography on new developments at international level, related to the use of Bots in medical issues, but more specifically those used against smoking.

5.3 Future lines of research

The use of a Bot to improve the impact of smoking cessation treatment represents a novel system that allows easier management of treatment administration; however, there is no evidence strong enough to argue that it is a successful way to achieve favorable results. This poses a challenge for future applications in which the implementation of a Bot is used as a reference to support these tasks, since they will have to comply with standards and guidelines established by specialized and traditionalist entities that often do not give this type of technology a chance.

It should not be forgotten that this type of mobile applications are considered inexpensive, upgradeable and accessible to most people who own a smartphone, so the development of new versions should obey the following lines of future research:

1. Development of new user-centered applications, but based on scientific evidence of treatments.
2. Extend developments to new platforms to make them more accessible to users.
3. Implement within the Bot programming, means of communication by natural language, such as voice recognition.
4. Training for health professionals so that they see the tool as a support and not as a limitation.
5. Strengthen the methods for measuring the impact on the effectiveness of treatments, in order to have a solid basis on the use of this type of applications.

Bibliography

Adenowo, A. A., & Adenowo, B. A. (2013). Software engineering methodologies: A review of the waterfall model and object-oriented approach. *International Journal of Scientific & Engineering Research*, *4*(7), 427–434.

Alemán Cortes, A. V. (2017). *Mobile application for the prevention of addictions in the context of the UCLV* [PhD Thesis]. Central University "Marta Abreu" of Las Villas.

Alfonseca, M. (2014). *is the Turing test sufficient to define "artificial intelligence"?*

Anderson, P., & Montero, B. (2021). *Mobile application for oral health promotion* [B.S. thesis]. National University of Chimborazo.

Arias, D., & Vela, H. (2015). Application of color theory and responsive web design techniques in front-end application development. *en. In: pg, .77*

Arias Gonzáles, J. L., and Covinos Gallardo, M. (2021). *Research design and methodology*.

Ávila-Tomás, J. F., Espinosa, E. O., Lorenzo, C. M., Suberviola, F. J. M., Pardo, B. M., Serrano, M. E. S., Güeto-Rubio, M. V., & Dej, G. (2020). Dejal@ Bot: A chatbot applicable in the treatment of smoking cessation. *Journal of Research and Education in Health Sciences (RIECS)*, *5*(1), 33–41.

Bacilio Ruiz, A. (2021). *Evaluation of the use of a Chatbot for follow-up in a clinical trial of COVID-19 prophylaxis in healthcare personnel.*

Baena Paz, G. (2017). *Metodología de la investigación*. Grupo editorial patria.

Balaguera, Y. D. A. (2013). Agile methodologies in the development of applications for mobile devices. Current status. *Technology Journal, 12*(2), 111–123.

Barker Maillard, C. (2019). *Evaluation and development of virtual assistance system on sexuality based on artificial intelligence tools to support educational work in schoolchildren.*

Benito Rodríguez, M. (2018). *File repository and storage in Telegram.*

Betjeman, T. J., Soghoian, S. E., & Foran, M. P. (2013). MHealth in sub-Saharan Africa. *International journal of telemedicine and applications, .2013*

Bistarelli, S., Fioravanti, F., Peretti, P., & Santini, F. (2012). Evaluation of complex security scenarios using defense trees and economic indexes. *Journal of Experimental & Theoretical Artificial Intelligence*, *24*(2), 161–192.

Blanco, A., Sandoval, R. C., Martínez-López, L., & Caixeta, R. de B. (2017). Ten years of the WHO Framework Convention on Tobacco Control: Progress in the Americas. *mexico public health*, *59*, 117–125.

Bonet, L., Izquierdo, C., Escartí, M. J., Sancho, J. V., Arce, D., Blanquer, I., & Sanjuan, J. (2017). Use of mobile technologies in patients with psychosis: A systematic review. *Journal of Psychiatry and Mental Health*, *10*(3), 168–178.

Bourke, J., Kirby, A., & Doran, J. (2016). *Survey & Questionnaire Design*. Cork, IRL: NuBooks.

Briz Ponce, L. (2016). *Analysis of the effectiveness in m-health Applications on mobile devices within the field of medical training.*

Caballero, A. (2014). *Innovative comprehensive methodology for plans and theses*. CENGAGE Learning.

Caballo Trebol, Á. (2013). Measuring credit risk: Development of a new tool. *Credit risk measurement.*, 1–195.

Cabrera Mendoza, N. I., Castro Enriquez, P. P., Demeneghi Marini, V. P., Fernández Luque, L., Morales Romero, J., Sainz Vazquez, L., & Ortiz León, M. C. (2014). mSalUV: A new mobile messaging system for diabetes management in Mexico. *Pan American Journal of Public Health*, *35*, 371–377.

Cahn, J. (2017). CHATBOT: Architecture, design, & development. *University of Pennsylvania School of Engineering and Applied Science Department of Computer and Information Science*.

Cancio, L. P., & Bergues, M. M. (2013). Web site usability, methods and techniques for evaluation. *Revista Cubana de Información en Ciencias de la Salud,24* (2), Art. 2. http://www.rcics.sld.cu/index.php/acimed/article/view/405

Centros de Integración Juvenil | Gobierno | gob.mx. (2022, January 1). https://www.gob.mx/salud%7Ccij/que-hacemos

Chavez Vizuet, Psic. e. (2016). *Smoking Cessation Treatment Juvenile Integration Centers Treatment and Rehabilitation Directorate Application Manual.* http://www.intranet.cij.gob.mx/Archivos/Pdf/MaterialDidacticoTratamiento/ManualTerapiaParaDejardeFumar.pdf

Chesñevar, C. I., & Estevez, E. C. (2018). *E-commerce in the age of bots*.

Contreras, H. (2012). *Theory of Computing for Systems Engineering: A practical approach*. Retrieved from: http://webdelprofesor. ula. ve/ingenieria/hyelitza/materias ....

Crismán-Pérez, R., & Núñez-Vázquez, I. (2015). Morphological knowledge scale of the Andalusian linguistic modality: Reliability studies, validity evidence and didactic implications. *Revista de Investigación Lingüística*, *18*, 43–64.

Cruz Barrera, D. A., & Zambrano Lazarte, N. F. (2020). *Chatbot for learning about sexuality*.

Cruz Zapata, B. (2018). A proposal for usability requirements knowledge management in mobile health applications. *Research project:*

Dahiya, M. (2017). A tool of conversation: chatbot. *International Journal of Computer Sciences and Engineering*, *5*(5), 158–161.

Diaz Guerra, M. E. (2021). *Chatbot for learning breast cancer prevention*.

Estrada Cutimbo, L. (2018). *Implementing artificial intelligence-based chatbot for requirements and incident management in an insurance company*.

Fernández, A. M. (2020). *Health-related mobile applications. A study on applications with functionality for medication intake reminder* [PhD Thesis]. University of Zaragoza.

Formella, A. (2010). Automata theory and formal languages. *Department of Computer Science, Universidade de Vigo, June*.

Fred, K., & Howard, L. (2002). *Investigacion del Comportamiento Metodos de Investigacion Ciencias Sociales*. Mexico: McGRAW-Hill/interamericana editores, sa de cv.

Freire, C. E. E. E. E. (2018). Variables and their operationalization in educational research. Part I. *Conrado Journal*, *14*(65), 39–49.

Freire, E. E. E. E. (2019). Variables and their operationalization in educational research. Second part. *Conrado Journal*, *15*(69), 171–180.

Frías-Navarro, D. (2014). Apuntes de SPSS. *University of valencia.*, 1–10.

Frías-Navarro, D. (2022). Notes on estimating the internal consistency reliability of the items of a measurement instrument. *D. Frías-Navarro, Recommendations for writing a research report and critical reading. Spain: University of Valencia. Retrieved from https://www. uv. es/friasnav/AlfaCronbach. pdf.*

Galán Pache, L. (2014). *Development of an interface for voice control of the mobile messaging application Telegram*.

García-Pazo, P., Fornés-Vives, J., Sesé, A., & Pérez-Pareja, F. J. (2020). Apps for smoking cessation using Cognitive Behavioral Therapy. A systematic review. *addictions*, *33*(4), 333–344.

Gil, A. M. C., & Afrashtehfar, K. I. (2020). Telegram Messenger: A suitable tool for Teledentistry. *Journal of Oral Research*, *9*(1), 4–6.

Gliem, J. A., & Gliem, R. R. (2003). *Calculating, interpreting, and reporting Cronbach's alpha reliability coefficient for Likert-type scales.*

González, A. B. G., & Reboredo, A. de L. (2019). *Web usability.*

González Alonso, J., and Pazmiño Santacruz, M. (2015). Calculation and interpretation of Cronbach's Alpha for the case of validation of the internal consistency of a questionnaire, with two possible Likert-type scales. *Journal publishing*, *2*(1), 62–67.

González Duque, R. (2011). *Python for everyone.* Creative Commons Attribution.

Greenwald, R., Stackowiak, R., & Stern, J. (2013). *Oracle essentials: Oracle database 12c.* O'Reilly Media, Inc.

Grimaldo Botero, G. J. (2013). *Development of a mobile application to support the web platform of the observatory "Monitoring of physical and physiological variables in children and adolescents of school age in Risaralda".*

Guerrero, J. S. D., Bazan, Y. Y. L., & Moreno, F. J. S. (2017). Chatbot development using Microsoft bot framework. *Espirales multidisciplinary research journal*, *1*(11).

Gutiérrez López, A., and Castillo Franco, P. (2008). Epidemiological study of alcohol and tobacco use in patients seeking treatment in CIJ in 2007. *Centros de Integración Juvenil, Dirección de Investigación y Enseñanza, Subdirección de Investigación, Informe de Investigación*, *8*, 12.

Hernández, R., Fernández, C., & Baptista, P. (2016). Metodología de la investigación. 6th Edition Sampieri. *Soriano, RR (1991). Guía para realizar investigaciones sociales. Plaza y Valdés.*

Hernández-Sampieri, R., Fernández Collado, C., & Baptista Lucio, P. (2018). *Research methodology* (Vol. 4). McGraw-Hill Interamericana Mexico.

Hernández-Sampieri, R., & Mendoza Torres, C. P. (2018). *Research methodology: Quantitative, qualitative and mixed routes*. McGraw Hill Mexico.

Hutton, B., Catalá-López, F., & Moher, D. (2016). The PRISMA statement extension for systematic reviews incorporating network meta-analyses: PRISMA-NMA. *Clinical Medicine*, *147*(6), 262–266.

Instituto Nacional de Psiquiatría Ramón de la Fuente Muñiz, S. de S. (2017). *Encuesta Nacional de Consumo de Drogas, Alcohol y Tabaco 2016-2017: Tobacco Report*. INPRFM Mexico City.

Kerlinger, F. N. (1979). *Behavioral research a conceptual approach*.

Kuri-Morales, P. A., González-Roldán, J. F., Hoy, M. J., & Cortés-Ramírez, M. (2006). Epidemiology of smoking in Mexico. *public health of mexico,48* , s91-s98.

Labra Chino, E., & Quispe Poma, E. (2022). *Referral method for chatbot-based health consultation care of older adult users in a pandemic context*.

Legaspi, L. M. D., Ramírez, L. C., & Olalde, M. G. C. (2020). Risk perception and alcohol and tobacco use among high school students in Zacatecas . *Lux Médica*, *15*(43), 13–24.

León, C., Xochilt, D., Cruz, A., & Betzaida, S. (2015). *Mobile technology applications as support for smoking cessation* (pp. 1998-2002).

GENERAL LAW FOR TOBACCO CONTROL, núm. DOF 06-01-2010, CHAMBER OF DEputies OF THE H. CONGRESS OF THE UNION (2010). http://www.conadic.salud.gob.mx/pdfs/ley_general_tabaco.pdf

Leyva-Vázquez, M., & Smarandache, F. (2018). *Artificial Intelligence: Challenges, perspectives and role of Neutrosophy*. Infinite Study.

Londoño Pérez, C., Rodríguez Rodríguez, I., & Gantiva Díaz, C. A. (2011). Questionnaire for the classification of cigarette consumers (C4) for young people. *Diversitas: perspectives in psychology*, *7*(2), 281–291.

López-Roldán, P., & Fachelli, S. (2015). *Quantitative social research methodology*.

Maida, E. G., and Pacienzia, J. (2015). *Software development methodologies*.

Mantilla, M. C. G., Ariza, L. L. C., & Delgado, B. M. (2014). Methodology for the development of mobile applications. *Tecnura: Technology and Culture Affirming Knowledge*, *18*(40), 20–35.

Martínez, C. M., & Sepúlveda, M. A. R. (2012). Introduction to exploratory factor analysis. *Colombian journal of psychiatry*, *41*(1), 197–207.

Matas, A. (2018). Likert-type scale format design: A state of the art. *Electronic journal of educational research*, *20*(1), 38–47.

Mazera, L., & Gonzalez, M. J. S. (2018). Bot of health care: development of the mobile application "Kiga" for chronic kidney disease in Brazil. *Qualitative research in digital communication and society: new challenges and opportunities*.

Medina Martínez, N. F. (2015). Las variables complejas en investigaciones pedagógica-Complex variables in pedagogical research. *Revista UPEU - Revista de Investigación Apuntes Universitarios*.

Mertens, D. M. (2019). *Research and evaluation in education and psychology: Integrating diversity with quantitative, qualitative, and mixed methods*. Sage publications.

MEX-ALVAREZ, D. C., HERNÁNDEZ-CRUZ, L. M., UC-RIOS, C. E., & CAB-CHAN, J. R. (n/d). Web usability analysis through standardized metrics and its practical application on the SAEFI platform Web usability analysis through standardized metrics and its practical application on the SAEFI platform. *Journal of*, 15.

Miranda Castellón, D. (2016). *Color Theory as an indispensable element within Graphic Design and Visual Communication.*

Montilva, J., Arapé, N., & Colmenares, J. (2003). *Component-based software development*. Proceedings of the IV. Automation and Control Congress. Mérida, Venezuela.

Mulyanto, A. D. (2020). Pemanfaatan Bot Telegram Untuk Media Informasi Penelitian. *MATICS*, *12*(1), 49–54.

Oberti, A., & Bacci, C. (2016). *Research Methodology.*

Olson, C., & Kemery, K. (2019). *Voice report From answers to action: Customer adoption of voice technology and digital assistants*. Microsoft.

Oviedo, H. C., and Campo-Arias, A. (2005). Approach to the use of Cronbach's alpha coefficient. *Colombian journal of psychiatry*, *34*(4), 572–580.

Pacheco Campoverde, L. G., and Idrovo Tapia, C. I. (2014). *Development of a mobile application in Android to support relapse prevention in patients in recovery process of the Psychiatric Hospital Humberto Ugalde Camacho* [B.S. thesis].

Pedroza, H., and Dicovskyi, L. (2007). *Statistical analysis system with SPSS.*

Pérez Peña, J. B., and Ramos Jurado, J. R. (2021). *Chatbot with artificial intelligence for the customer service process in the urology service of a health care facility.*

Pericot Valverde, I. (2016). *Applications of virtual reality in the treatment of smoking.*

Pressman, R. S. (2010). *Software Engineering A Practical Approach* (Seventh Edition). McGraw-Hill.

Puerto, G. A. S., Rincón, E. H. H., & Obando, F. S. (2016). Mobile health apps: Use in Internal Medicine patients at the Regional Hospital of Duitama, Boyacá, Colombia. *Cuban Journal of Health Sciences Information (ACIMED)*, *27*(3), 271–285.

Pulido, J. F. G. (2018). Construct validation to a questionnaire related to the strategic diagnosis of ICT in Higher Education. Case study. *Acción Pedagógica*, *27*(1), 22–33.

Rabelo, R. J., Romero, D., & Zambiasi, S. P. (2018). Softbots supporting the operator 4.0 at smart factory environments. *IFIP International Conference on Advances in Production Management Systems.*, 456–464.

Rey Iborra, C. (2019). *Mobile health applications as tools to support chronic patient self-management of care* [B.S. thesis].

Rodrigues, J. L. T., Alves, C. F., & Osshiro, M. (n/d). *Anetha-Desenvolvimento de Chatbot em Python*.

Rodríguez, D. R. (2022). *Creation of chatbot for canine veterinary consultation using artificial intelligence*.

Rouhiainen, L. (2018). Inteligencia artificial. *Madrid: Alienta Editorial*.

Rowland, S. P., Fitzgerald, J. E., Holme, T., Powell, J., & McGregor, A. (2020). What is the clinical value of mHealth for patients? *NPJ digital medicine*, *3*(1), 1–6.

Ruiz, E. F., Proaño, Á., Ponce, O. J., & Curioso, W. H. (2015). Mobile technologies for public health in Peru: Lessons learned. *Peruvian Journal of Experimental Medicine and Public Health*,*32* (2), Art. 2.

Segrelles-Calvo, G., de Granda-Beltrán, A. M., & de Granda-Orive, J. I. (2021). A chatbot for smoking cessation. will it be the future? *addictions*, *33*(1), 73–74.

Sekulovski, E. (2014). *Cross-Platform Mobile Application Generation with a Model-Driven Approach Based on IFML and Cross-Compilation*. Scuola di Ingegneria dell'Informazione.

Serrano Mascaraque, E. (2009). Accessibility vs. web usability: Evaluation and correlation. *Librarianship research*, *23*(48), 61–103.

Solís, G. A., Méndez, G., & Segura, R. J. (2014). *SOFTWARE TESTING FOR ANDROID MOBILE DEVICES*.

Sommerville, I. (2005). *Software engineering*. Pearson Education.

Suárez, O. M. (2007). Application of factor analysis to market research. Case study. *Scientia et technica*, *1*(35).

Telegram vs Whatsapp: Which is the best messaging app [December 2021] (n/d). GEEKNETIC. Retrieved December 9, 2021, from https://acf.geeknetic.es/Guia/1839/Telegram-vs-Whatsapp-Cual-es-la-mejor-app-de-mensajeria.html.

Velázquez Macías, J., Vela Dávila, J., Veyna Lamas, M., & Gomez Aguilar, C. (2017). Development of a bot for support in the treatment of smoking in the Juvenile Integration Center in Zacatecas. *Information Technology and Communications Journals*, *1*(2).

Velázquez Macías, J., Veyna Lamas, M., Vela Dávila, J., & Rodríguez González, B. (2016). Use of a Bot for the verification of mathematical formulas of the subjects of Probability and Operations Research at the Polytechnic University of Zacatecas. *Journal of Computer Systems and ICT's*, *2*(5), 53–58.

Velázquez-Altamirano, M., and Córdova-Alcaráz, A. J. (2020). *INSTITUTIONAL SYSTEM FOR THE EVALUATION OF TREATMENT PROGRAMS.*. 23.

Velázquez-Macias, J., Vela-Dávila, J. A., Veyna-Lamas, M., & Pinales-González, L. C. (2017). Development of a mobile application as a support in the prevention of type 2 diabetes in people over 18 years old. *Information and Communications Technology Journals*, *1*(2), 47–55.

Velázquez-Macias, J., Veyna-Lamas, M., Vela-Dávila, J. A., Lara-Torres, C. G., & González-Saenz, J. A. (2020). Analysis of the risk of type 2 Diabetes by means of the Diabetest mobile application in people over 18 years old in the municipality of Fresnillo, Zacatecas, Mexico. *Journal of Mathematical Programming and Software*.

Villegas-Ch, W., Arias-Navarrete, A., & Palacios-Pacheco, X. (2020). Proposal of an Architecture for the Integration of a Chatbot with Artificial Intelligence in a Smart Campus for the Improvement of Learning. *Sustainability*, *12*(4), 1500.

Vique, R. R. (2019). *Methods for mobile application development.*

Vogt, W. P., & Johnson, R. B. (2015). *The SAGE dictionary of statistics & methodology: A nontechnical guide for the social sciences*. Sage publications.

VonHoltz, L. A. H., Hypolite, K. A., Carr, B. G., Shofer, F. S., Winston, F. K., Hanson III, C. W., & Merchant, R. M. (2015). Use of mobile apps: A patient-centered approach. *Academic Emergency Medicine*, *22*(6), 765–768.

WHO (2015). *WHO global report on trends in prevalence of tobacco smoking 2015*. World Health Organization.

Yoong, S. L., D'Espaignet, T. E., Wiggers, J., St Claire, S., Mellin-Olsen, J., & Grady, A. (2020). *WHO summaries of tobacco literacy: Tobacco and postoperative complications*.

Zavala-Arciniega, L., Fleischer, N., Paz-Ballesteros, W. C., Reynales-Shigematsu, L. M., Meza, R., & Jimenez-Mendoza, E. (2019). Challenges in the implementation of MPOWER Measures in Mexico. Results from the Global Adult Tobacco Survey (GATS) 2009-2015. *APHA's 2019 Annual Meeting and Expo (Nov. 2-Nov. 6).*

Annexes

Annex 1

TOBACCO USE MOTIVES QUESTIONNAIRE (FINAL)

Institution :________________ Date: ______/______/______
Day Month Year

Nombre:___________________________________		
Father's last name Mother's last name First name(s)		
Gender:		Age:

Below is a list of reasons why some people continue to smoke. Read each of the reasons and respond according to your own experience.
Mark with an "X":
1) Never
2) Rarely
3) Occasionally
4) Followed by
5) Frequently
6) Always

		1	2	3	4	5	6
1	I feel that smoking gives me security						
2	I feel that smoking makes me feel better						
3	Exhaling each puff of smoke gives me a pleasant sensation.						
4	I enjoy smoking after meals, with tea, coffee or alcohol.						
5	When I feel angry about something, I smoke to calm down.						
6	Even when sick I feel the need for a cigarette						
7	All the cigars I smoke are pleasurable.						
8	If I don't smoke, I lose part of my personality.						
9	I smoke to stay awake						
10	Feeling the cigar between your fingers is rewarding.						
11	When I am relaxed and in rest periods I like to smoke.						
12	If I'm nervous about something, I smoke almost twice as much.						
13	If I switch to mild cigars I smoke almost twice as much.						
14	I light a cigarette without having finished the previous one.						
15	I smoke more when I am at social gatherings						

16	In monotonous or boring jobs I smoke more.						
17	I enjoy hitting the cigar in a certain way.						
18	During my work, I take time to enjoy a cigar.						
19	I smoke when I want to forget my worries						
20	I feel unpleasant if I don't smoke, even though smoking is not really pleasant.						
21	I light cigarettes without realizing it						
22	I smoke so as not to feel so lonely.						
23	I feel more alert and energetic when smoking cigarettes.						
24	I enjoy smoking from the moment I hold the pack in my hands.						
25	When I am quiet I like to smoke						
26	I smoke more when I am tense.						
27	When I stop smoking for a few hours, I begin to feel unpleasant physical symptoms.						
28	There have been times when I forget where I left a cigarette burning.						
29	Having a cigarette in my bag gives me peace of mind.						
30	I work better when I smoke						
31	Taking out the pack, holding the cigarette in your hand and seeing the smoke, is gratifying.						
32	My desire to smoke increases when I am comfortable						
33	Smoking reduces my blood pressure						
34	When I don't have cigarettes, I'll do anything to get them.						
35	I smoke all the cigarettes offered to me						

Annex 2

BOT USABILITY QUESTIONNAIRE

Institution : _______________ Date: ______/______/______
Day Month Year

Nombre:____			
______________ Father's last name Mother's last name First name(s)			
Gender:		Age:	

Below is a list of questions related to the Bot management implemented in smoking cessation treatment. Read each of them and answer according to your own experience.
Mark with an "X":
1) Lousy
2) Bad
3) Regular
4) Good
5) Very good
6) Excellent

1	Installation of the Telegram program	1	2	3	4	5	6
2	The search for the contact "Quit smoking".	1	2	3	4	5	6
3	The Bot's help	1	2	3	4	5	6
4	Response time	1	2	3	4	5	6
5	The clarity of the available options	1	2	3	4	5	6
6	The clarity of the answers received	1	2	3	4	5	6
7	Ease of use	1	2	3	4	5	6
8	The graphical interface	1	2	3	4	5	6
9	Text size	1	2	3	4	5	6
10	Tips received (thematic)	1	2	3	4	5	6
11	Periodicity of tips received (time)	1	2	3	4	5	6
12	Data consumption	1	2	3	4	5	6
13	Overall speed of execution	1	2	3	4	5	6

Annex 3

TREATMENT EFFICACY

Institution :_______________ Date: ______/______/______
Day Month Year

Nombre:____		
______________________ Father's last name Mother's last name First name(s)		
Gender:		Age:

Intended for Therapists
Below is a list of questions related to the results obtained after the implementation of the bot in support of smoking treatment in some patients. Read each of them and answer according to your own experience in relation to the traditional method.
Mark with an "X" :
1) Lousy
2) Bad
3) Regular
4) Good
5) Very good
6) Excellent

		1	2	3	4	5	6
		1	2	3	4	5	6
		1	2	3	4	5	6
1	The reduction in consumption in units per day was	1	2	3	4	5	6
2	The reduction in treatment duration was	1	2	3	4	5	6
3	The use of time dedicated to treatment was	1	2	3	4	5	6
4	The reduction in incidences was	1	2	3	4	5	6
5	The increase in successful therapies was	1	2	3	4	5	6
6	The reduction of zero therapies was	1	2	3	4	5	6
7	The reduction of logbook losses was	1	2	3	4	5	6
8	The reduction of incorrect filling of logbooks was	1	2	3	4	5	6
9	In general, what was the patient's progress with this treatment?						

Annex 4

USABILITY OF THE WEB PLATFORM

Institution :_______________ Date: _____/_____/_____
Day Month Year

Nombre:____			
____________________ Father's last name Mother's last name First name(s)			
Gender:		Age:	

Intended for Therapists (5)
Below is a list of questions related to the management of the WEB application that manages the data provided by patients through the Bot in the smoking cessation treatment. Read each one of them and answer according to your own experience.
Mark with an "X":
1) I do not agree at all
2) I do not agree
3) Somewhat disagree
4) Some agreement
5) Agreed
6) I strongly agree

		1	2	3	4	5	6
1	I think I will enjoy visiting this website frequently.						
2	I found the website not complex						
3	It is relatively easy to use the website						
4	I do not need the support of an expert to browse the website.						
5	I found the different possibilities of the website to be quite well integrated.						
6	There is consistency in the website						
7	I imagine that most people would learn very quickly how to use the website.						
8	I found the website visually appealing						
9	I felt very confident in the handling of the website.						
10	I don't need to learn a lot of things before I can use the website.						
		1	2	3	4	5	6

Appendix A. Source code of the consumption records module

```
# -+- coding: utf-8 -*-
#tabaquismo2.py
import math


####################Declaracion de funciones


def fraseinicio():
      frases = array(["Estas seguro de continuar con un ","Que tal sigamos con un","\xF0\x9F\x98\xA3 ","No quería hacerlo pero sigamos con un ","Bien comencemos con un", "Ok comencemos con un", "No esperaba que seleccionaras esto pero continuemos", "Complicado pero continuemos con un "])

      x=random.randrange(0,8)
      frase=frases[x]

      return frase


################Fin declaración de funciones


import telepot, time, pprint, random, numpy
from numpy import array
print ('Escuchando conversaciones....')

def handle(msg):
      chat_id = msg['chat']['id']
      print chat_id
      command = msg['text']
      desde= msg['chat']['first_name'] + ' ' + msg['chat']['last_name'] + ' - Tabaquismo'

      print (desde)
      print ('Comando recibido: %s' % command)

    #####Evaluando valores para /tc #####
```

```
lenop=len(command)
      if lenop>3:
                  op=command[0]+command[1]+command[2]
                  print (op)
      elif lenop>2:
                  op=command[0]+command[1]
                  print (op)
      elif lenop>1:
                  op=command[0]
                  print (op)
      elif lenop<=1:
                  op=command[0]
                  print (op)
      #/xx 1,2,3

      if op == '/co':
            ini=0
            fi=len(command)

            inv1=4                              #inicio cadena 1
            finv1=command.find(",")                #fin cadena 1
            v1=""
            for x in range(4, finv1):
                  v1=v1+command[x]

            print ("motivo: " + str(v1)) #v1 obtenido
            ####################################################

            inv2=finv1+1                              #inicio cadena 2
            finv2=command.find(",",inv2,len(command))          #fin cadena 2
            v2=""
            for x in range(inv2, finv2):
                  v2=v2+command[x]

            print ("numero 2: " + str(v2)) #v2 obtenido
            ####################################################
```

```
for x in range(inv3, finv3):
        v3=v3+command[x]

    print ("numero 3: " + str(v3)) #v3 obtenido

    ###################################################

    inv4=finv3+1                        #inicio cadena 4
    finv4=len(command)                 #fin cadena 4
    v4=""
    for x in range(inv4, finv4):
        v4=v4+command[x]

    print ("numero 4: " + str(v4)) #v4 obtenido

    #Gestion de errores para /ci
    try:
        v1=int(v1)
        v2=str(v2)
        v3=str(v3)
        v3=str(v4)

        msgtemp="🆗 \xF0\x9F\x98\xA3" + "Registro correcto del
consumo de un cigarro por el motivo=" + str(v1)
        #llamada a funcion para almacenar
str(calculosPe(m,d,x,o))
        bot.sendMessage(chat_id, msgtemp)
        #####Fin Evaluando valores para /ci #####

    except ValueError:
        msgtemp="😓" + " Algo salio mal con los valores que
recibi, ¿podrias intentarlo de nuevo?"
        bot.sendMessage(chat_id, msgtemp)
        #print( "☹" + " Algo salio mal con los valores que
recibi, ¿podrias intentarlo de nuevo?")

elif command == '/start':
```

```
bot.sendMessage(chat_id, "Puedes utilizar los siguientes comandos para acceder a cada accion.")

            bot.sendMessage(chat_id, "/cigarro - Registrar un nuevo consumo\n/motivo - Muestra los posibles motivos por los que se puede dar del consumo \n/cuestionario - Realizar el cuestionario de motivos de consumo de tabaco (consultar los resultados con su terapeuta)\n/ayuda - accede a la ayuda ")


      elif command == '/cigarro':

            bot.sendMessage(chat_id, fraseinicio()+ "registro de consumo de un cigarro. Teclea el comando /co seguido del numero del motivo por el que fumas un cigarrillo en este momento, el lugar, la actividad que estas realizando y el sentimiento percibido.\n\nEjemplo: /co 2,mi trabajo,hora de descanso,tristeza")

      elif command == '/motivos':

            bot.sendMessage(chat_id, "A continuación se describen algunos de los motivos por los cuales fumas:\nMotivo 1 = En compañia de personas\nMotivo 2 = Para concentrarme mejor y evitar la fatiga cuando trabajo\nMotivo 3 = Siento agradables los movimientos del fumar y ver el humo como se esparce\nMotivo 4 = Con el café, después después de los alimentos o en periodos de descanso.\nMotivo 5 = Cuándo me siento tenso enojado o preocupado.\nMotivo 6 = Al no fumar, por más de 30 minutos me siento mal y las molestias se quitan al fumar.\nMotivo 7 = Fumo por placer.\nMotivo 5 = No me percato cuando enciendo el cigarro.")

      elif command == '/cuestionario':

            bot.sendMessage(chat_id, "Has iniciado el Cuestionario de Motivos de Consumo de Tabaco a continuación se presenta una lista de motivos por los cuales algunas personas continuan fumando. Lee cada uno de los motivos y responde de acuerdo a tu propia experiencia.\nContesta con numeros si:\n3) Si tu 'Muy frecuentemente' fumas por ese motivo.\n2) Si tu 'ocasionalmente' fumas por ese motivo.\n1) Si tu 'nunca' fumas por ese motivo.\n¿Estas seguro(a) de continuar? s/n  ")

      elif command == '/ayuda':

            bot.sendMessage(chat_id, "/cigarro - Registro de consumo de un cigarro \n/motivos - accede a la lista de motivos por los cuales fumas")

      elif command == 'Gracias' or command == 'gracias':

            bot.sendMessage(chat_id, "👀 De nada...")

      elif command == 'Hola' or command == 'hola':

            bot.sendMessage(chat_id, "✌ Hola, que tal")

      elif command == 'buenos dias' or command == 'Buenos dias':

            bot.sendMessage(chat_id, "☀ Buenos dias...")

      elif command == 'buenas noches' or command == 'Buenas noches':

            bot.sendMessage(chat_id, "🌙 Buenas noches...")

      elif command == 'buenas tardes' or command == 'Buenas tardes':

            bot.sendMessage(chat_id, "☁ Buenas tardes...")
```

```
bot.sendMessage(chat_id, "😫 Bye Bye...")
      elif command == 'Como estas' or command == 'como estas' or
command == 'Kmo estas' or command == 'kmo estas':
            bot.sendMessage(chat_id, "😉 Muy Bien, ¿Y tu? ")

      elif command == 'Mal' or command == 'mal' :
            bot.sendMessage(chat_id, "😟 No me digas eso")
      elif command == 'Gracias' or command == 'gracias' :
            bot.sendMessage(chat_id, "😉 De nada")

      elif command == 'Bien' or command == 'bien' or command == 'Bien
tambien' or command == 'bien tambien' or command == 'bien gracias' or
command == 'Bien tambien':
            bot.sendMessage(chat_id, "😉 Genial")
      else:
            bot.sendMessage(chat_id, "😟" + "    No reconozco el
comando, ¿Podrias intentar de nuevo? o consulta la /ayuda")



# Create a bot object with API key
#bot = telepot.Bot('222843035:AAHxUCQ_dSLogfUgULWg4wnoH-AQu9nzM94')
#formulas
bot              =              telepot.Bot('449353037:AAFCFjElR5zt6M6Nn-
srFVxoKpCrPX_2apM')#Dejardefumar


# Attach a function to notifyOnMessage call back
#bot.notifyOnMessage(handle)
bot.message_loop(handle)


# Listen to the messages
while 1:
    time.sleep(180)
    #bot.sendMessage(233304165, "😉 Genial")
```

MIX
Papier aus verantwortungsvollen Quellen
Paper from responsible sources
FSC® C105338

Printed by Books on Demand GmbH, Norderstedt / Germany